British Quaternary Studies

RECENT ADVANCES

British Quaternary Studies

RECENT ADVANCES

EDITED BY

F.W. SHOTTON

Emeritus Professor of Geology
University of Birmingham

CLARENDON PRESS · OXFORD

1977

Oxford University Press, Walton Street, Oxford OX2 6DP

OXFORD LONDON GLASGOW NEW YORK
TORONTO MELBOURNE WELLINGTON CAPE TOWN
IBADAN NAIROBI DAR ES SALAAM LUSAKA ADDIS ABABA
KUALA LUMPUR SINGAPORE JAKARTA HONG KONG TOKYO
DELHI BOMBAY CALCUTTA MADRAS KARACHI

© Oxford University Press 1977

British Library Cataloguing in Publication Data

British Quaternary studies.
  1. Geology, Stratigraphic — Quaternary
  2. Geology, Great Britain
  I. Shotton, F W
  551.7'9'0941   QE696   77-30160

  ISBN 0-19-854414-6

Typeset by Hope Services, Wantage
Printed in Great Britain
by J.W. Arrowsmith Ltd., Bristol

This book is dedicated to the memory of

WALTER WILLIAM BISHOP

whose untimely death on 19 February 1977
deprived Quaternary science of an
imaginative researcher and an inspiring exponent

Preface

The Quaternary System covers approximately the last two million years of earth history and, if interest can be measured by the amount of study for each year of its duration, it is more intensively investigated than any other period of the past. The reason for this is not hard to find. During the Quaternary there were changes of climate, possibly of more dramatic rapidity and magnitude than ever before; we can study the shaping of the modern landscape or the migration of plants and animals as they reacted to the changing climate. The civil engineer meets many of his problems in Quaternary deposits. Above all, man must be fascinated by the opportunity to study the anatomy of his ancestors and their cultural development. So it is not surprising that the subject is truly interdisciplinary. It brings together for their mutual enlightenment the geologist, geomorphologist, botanist, zoologist, anatomist, palaeontologist, archaeologist, geophysicist, geochemist, climatologist, pedologist, soil mechanician, and increasingly nowadays, the computer scientist.

The International Union of Quaternary Research (INQUA) is the body which integrates these many interests and on the occasion of its holding its tenth quadrennial congress in the United Kingdom it seemed appropriate to put on record some aspects of work on the Quaternary which is being done by scientists in this country. The choice of topics was not easy and the selection of authors even more difficult. In several cases, what was at first envisaged as a solo effort developed into a composite work. The editor can only hope that the selection will make interesting, stimulating, perhaps sometimes provocative reading. It cannot fail to bring out the multi-faceted aspect of Quaternary studies.

Birmingham
June 1976   F.W.S.

Contents

List of Contributors

P.H. BANHAM
Department of Geology, Bedford College, University of London.

H.J.B. BIRKS
The Botany School, University of Cambridge.

W.W. BISHOP
Department of Geology, Queen Mary College, University of London.

G.S. BOULTON
School of Environmental Sciences, University of East Anglia, Norwich.

V.N.D. CASTON
The British Petroleum Company Ltd., London.

J.A. CATT
Department of Pedology, Rothamsted Experimental Station, Harpenden, Hertfordshire.

K.M. CLAYTON
School of Environmental Sciences, University of East Anglia, Norwich.

G.R. COOPE
Department of Geological Sciences, University of Birmingham.

N.G.T. FANNIN
Institute of Geological Sciences, Continental Shelf Unit II, Edinburgh.

B.M. FUNNELL
School of Environmental Sciences, University of East Anglia, Norwich.

SIR HARRY GODWIN
Emeritus Professor of Botany, University of Cambridge.

A.S. JONES
School of Mathematics, University of East Anglia and Department of Mathematics, University of Melbourne, Australia.

M.J. KENNING
School of Environmental Sciences, University of East Anglia, Norwich.

M.P. KERNEY
Department of Geology, Imperial College of Science and Technology, University of London.

CUCHLAINE A.M. KING
Department of Geography, University of Nottingham.

H.H. LAMB
Climatic Research Unit, School of Environmental Sciences, University of East Anglia, Norwich.

I.N. McCAVE
School of Environmental Sciences, University of East Anglia, Norwich.

List of Contributors

G.F. MITCHELL
Trinity College, University of Dublin.
THEYA MOLLESON
Department of Anthropology, British Museum (Natural History), London.
P.E.P. NORTON
Department of Zoology, University of Glasgow.
N.J. SHACKLETON
Sub-Department of Quaternary Research, University of Cambridge.
F.W. SHOTTON
Emeritus Professor of Geology, University of Birmingham.
A.J. STUART
Museum of Zoology, University of Cambridge.
R.G. WEST
The Botany School, University of Cambridge.
P. WORSLEY
Department of Geography, University of Reading.
J.J. WYMER
Research Fellow in Archaeology, University of Chicago.

1

Oxygen isotope stratigraphy of the Middle Pleistocene

N. J. SHACKLETON

Abstract

Good oxygen isotope records through the Middle Pleistocene are now available for the three oceans and for a wide latitudinal range. These have provided a rigorously applicable stratigraphic scheme for Middle Pleistocene deep-sea deposits. A few biostratigraphic datums have been calibrated against the oxygen isotope stratigraphy (the extinction of *Stylatractus universus*, the extinction of *Pseudoemiliania lacunosa*, the extinction of *Globoquadrina pseudofoliata*), but there is no possibility of achieving the stratigraphic refinement which is needed for Pleistocene studies by biostratigraphic means alone.

Introduction

Oxygen isotope analysis was initiated in Cambridge by Dr. (now Sir Harry) Godwin, and the laboratory was set up by the writer. Since 1969 it has been the aim of successive projects, supported by NERC, to study the Upper Pleistocene, the Middle Pleistocene and, at present, the Lower Pleistocene. Oxygen isotope work on the Upper Pleistocene has recently been reviewed by the writer elsewhere (Shackleton 1977). Work on the Lower Pleistocene is in its infancy and only one good record has so far been obtained (Shackleton and Opdyke 1976). It therefore seems appropriate to concentrate in this review on the Middle Pleistocene, for much progress has been made in studies of this time period since it was reviewed by the writer in 1973 (Shackleton 1975).

Basic principles

There are three stable isotopes of oxygen (Table 1.1).

Table 1.1.
The abundance of the stable oxygen isotopes in air†

Mass Number	Abundance (per cent)
16	99·759
17	0·0374
18	0·2039

†From Nier (1950).

The work of Urey (1947) established the theoretical basis for the application to geological studies of measurements of the $^{18}O/^{16}O$ ratio in natural substances. In this major review Urey discussed, from a theoretical point of view, the distribution of isotopes between phases or compounds in equilibrium exchange reactions. One of the many cases that he discussed was the deposition of calcite from solution in water. He calculated that the isotopic fractionation factor specifying the difference in ^{18}O abundance between the carbonate ion and the water should be about 1·0220 at 0°C and about 1·0176 at 25°C. Initially it was the possibility of determining temperatures in the geological past, which was opened up by this calculation, that stimulated attempts to measure the ^{18}O content of carbonates with extreme precision. Epstein, Buchsbaum, Lowenstam, and Urey (1953) established an empirical temperature-estimating equation using carbonate deposited by molluscs, and only two years later Emiliani (1955) published a now-classic paper in which he showed oxygen isotope analyses of foraminiferal tests from deep-sea cores in the Caribbean, Atlantic, and Pacific.

Oxygen isotope measurements are published using the δ notation, where δ is defined by

$$\delta = 1000 \left(\frac{{}^{18}O/{}^{16}O \text{ sample}}{{}^{18}O/{}^{16}O \text{ standard}} - 1 \right)$$

and hence is expressed in parts per thousand ('per mil'), written ‰.

Emiliani (1955) discussed the principles and perils of oxygen isotope palaeotemperature analysis in some detail. Bowen (1966) reviewed the whole field, and the reviews of Thorley (1961) and Hecht (1976) cover aspects of it in more detail.

Since 1955 the main development in work on Pleistocene deep-sea sediments has been the appreciation that variations in ocean ^{18}O content play a greater role than Emiliani originally realized. During each glacial, isotopically light ice accumulated on the continents, leaving the oceans slightly enriched in ^{18}O. Consideration of the known isotopic composition in snow over the Antarctic ice-sheet led Olausson (1965) to suggest that the effect on oceanic isotopic composition was closer to 1·0 ‰ than to the 0·4 ‰ estimated by Emiliani. Dansgaard and Tauber (1969) re-emphasized this point, and Shackleton (1967) confirmed it directly by making oxygen isotope analyses of benthonic Foraminifera from deep water, where the effect of changes in temperature must have been small or negligible. The conclusion of that paper was that the chief value of oxygen isotope sequences in Pleistocene deep-sea sediment cores lies in their direct and continuous record of changing continental ice volume (and of its inverse, the volume of oceanic water and hence sea level). There are two aspects of this side of oxygen isotope work that will be reviewed here. First, it is of interest to derive the best possible record of actual changes in ice volume as a function of time. Such an ideal record will presumably be based on measurements from several cores. Secondly, the observationof this record, albeit distorted, in a particular core of sediment permits sections of the core to be correlated with the standard sequence and hence dated. Thus oxygen isotope analysis has become both the means of obtaining one type of standard glacial stratigraphic sequence, and the prime means of correlating to that sequence.

Materials analysed
Coccoliths
In principal, coccoliths might be superior to Foraminifera for oxygen isotope analysis because, since they are limited by their requirement

for light to living near the ocean surface, they are not subject to the depth-habitat uncertainty which complicates work with Foraminifera. Experimental studies using laboratory-cultured specimens have so far been pursued in at least two laboratories with rather discouraging results. On the other hand, analyses of pre-Pleistocene fossil material (Savin, Douglas, and Stehli 1975) has been more encouraging, and Margolis, Kroopnick, Goodney, Dudley, and Mahoney (1975) obtained a record from DSDP site 277 closely parallelling the one published for Foraminifera by Shackleton and Kennett (1975). Results from Pleistocene sediment cores are awaited with interest.

Shallow-living planktonic Foraminifera

Most of the samples analysed by Emiliani (1955) were of *Globigerinoides sacculifer*. Emiliani (1954) had previously shown by isotopic analysis that in recent sediments this species has an ^{18}O content consistent with growth in surface water. *G. ruber* was also analysed in two cores and was found to average 0·2 ‰ lighter, from which it was inferred that growth took place at a shallower depth. By analysing specimens of these two species collected together in shallow plankton tows, Shackleton, Wiseman, and Buckley (1973) concluded that the observed isotopic difference must be due to a departure from isotopic equilibrium, since the specimens they collected presumably lived at the same depth. This departure from isotopic equilibrium implies that the calcium carbonate was deposited under conditions other than those of thermodynamic equilibrium as specified in Urey (1947). Hence a species-dependent correction factor for departure from isotopic equilibrium must be applied before temperature is estimated from the isotopic composition of a Foraminiferal test. Vincent and Shackleton (in press) have concluded that over the Mozambique Channel area approximate summer isotherms can be drawn on isotopically determined temperatures from *G. sacculifer* in the 20–27°C temperature range. There is, however, no certainty that even this species is always represented in the sediment by a population whose members all deposited all their calcite at the ocean surface. In the Pacific cores that were analysed, the core top samples seem to represent temperatures from rather deeper than the surface (Shackleton and Opdyke 1973, 1976) and may therefore be subject to the uncertainties applying to the deeper-living Foraminifera described below.

Deeper-living planktonic Foraminifera

Emiliani (1955) and Lidz, Kehm, and Miller (1968) showed that deeper-living species such as *Globorotalia menardii* register lower isotopically derived temperatures than the *Globigerinoides* species and also exhibit a smaller isotopic difference between glacial and interglacial specimens. By assuming that the species occupies a constant depth habitat, they argued that the observed glacial–interglacial isotopic range set an upper limit on the glacial–interglacial change in ocean isotopic composition. Shackleton (1968) criticized this argument on the basis that the habitat of these species is more likely to be related to constant water density than to a constant depth; during glacials the ocean salinity was increased, so that the same density would be found at a higher temperature and a different depth. Thus as ice sheets accumulated and the density of the oceans became gradually greater, the species were probably forced to live in water of higher and higher temperature. Shackleton and Vincent (in preparation) have confirmed this by showing that specimens of *G. menardii* from the Atlantic, with its high salinity, register a higher isotopic temperature than those from the less saline Indian Ocean. Savin and Douglas (1973) suggested that species adapt to a particular osmotic pressure rather than a particular density, but this seems inherently unlikely to the author. An additional complication is provided by Grazzini's (in press) suggestion that *Globorotalia inflata* (and by implication, other deep-dwelling species) may depart from isotopic equilibrium. It may be concluded that the ocean isotopic record is not well recorded by deeper-living species.

Benthonic Foraminifera

It was argued by Shackleton (1967) that the temperature of the ocean deep-waters is unlikely to have changed much during the Pleistocene, since its minimum is controlled by the temperature of Antarctic Bottom Water, so that benthonics should yield the best isotopic records. However, Duplessy, Lalou, and Vinot (1970) have shown that there are departures from the oxygen isotopic equilibrium in benthonic as well as planktonic Foraminifera. In the core they studied, *Planulina wuellerstorfi* appeared about $1 \cdot 1 °/\text{oo}$ isotopically negative with respect to *Pyrgo* spp. More recently Duplessy, Chenouard. and Vila (1975) have shown in a core from the Norwegian Sea that, in addition to this difference, *Pyrgo* yields less consistent results, as has also been observed in the Cambridge laboratory. Shackleton and Opdyke (1973) obtained a record through the last 100 000 years using *Uvigerina, Pyrgo*, and *Planulina* by making an adjustment to correct for departure from equilibrium. Shackleton (1974) showed that *Uvigerina* calcite is at or very near isotopic equilibrium, and so far the most convincing benthonic data have been obtained from that genus.

Siliceous organisms

Labeyrie (1974) has established the feasibility of obtaining oxygen isotope records from siliceous diatoms. At the time of writing no core records have been published, but this appears to be a promising field.

Sources of material for ideal records

It may be concluded from the above discussion that the best records of changes in the oxygen isotopic composition of the oceans are likely to emerge from analyses of monospecific samples of benthonic Foraminifera. In principle sea surface temperatures would then be calculated by taking the difference between such a record and one obtained from coccoliths. Ideally the record should be from the deepest and coldest water, so as to minimize possible temperature effects, but one is limited to cores from above the calcite compensation depth. An additional limitation is imposed by the accumulation rate of sediments. The lower the accumulation rate, the more the record is blurred by the homogenizing effects of burrowing organisms. As regards oxygen isotope records, this problem was discussed by Shackleton and Opdyke (1973, 1976) and will be considered only briefly below. So far, the most detailed records have come from areas of upwelling off North America (Ninkovich and Shackleton 1975) and North Africa (Shackleton, in press). Almost equally good records, however, have been obtained by analysing *Globigerina bulloides* in Subantarctic cores from the Indian Ocean (Hays, Lozano, Shackleton, and Irving 1976; Hays, Imbrie, and Shackleton 1976). Unfortunately the cores from areas of upwelling suffer a special disadvantage. The high accumulation rate is due in large measure to the wind-induced upwelling of nutrient-rich water, and is thus climatically controlled. In both the areas mentioned, accumulation rates seem to have been substantially higher during glacial time.

The majority of oxygen isotope records extending into the Middle Pleistocene have been obtained in areas of slower sediment accumulation, between about 1 cm/1000 years and 2·5 cm/100 years, and are thus to a greater or lesser extent degraded by bioturbation. Emiliani (1958, 1964, 1966) has analysed several cores from the Caribbean region and parts of the North Atlantic, and has synthesized a generalized record for that area (Emiliani 1966). The Cambridge laboratory has concentrated on expanding the geographical range covered by samples. Shackleton and Opdyke (1973) analysed an exceptionally good core from the Western Equatorial Pacific (V28-238, Fig. 1.1). The importance of this publication was that it placed the oxygen isotope record within the framework of palaeo-

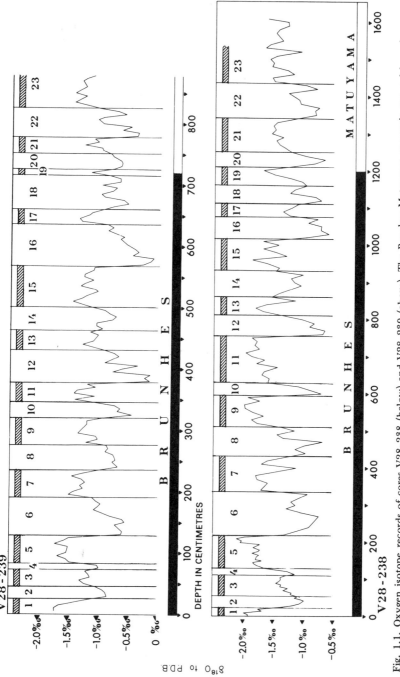

Fig. 1.1. Oxygen isotope records of cores V28–238 (below) and V28–239 (above). The Brunhes–Matuyama magnetic reversal boundary was located at 1200 cm in V28–238, and at 735 cm in V28–239. From Shackleton and Opdyke (1973 and 1976).

magnetic reversals and thus greatly strengthened the chronological side of deep-sea stratigraphy. Since that time, analyses have been carried out on several more cores from a wide geographical range and penetrating well into the Pleistocene. Fig. 1.1 shows part of a long record from the Western Equatorial Pacific extending into the Pliocene. Fig. 1.2 shows records from the North Pacific, the Eastern Equatorial Pacific, and the Equatorial and Subantarctic Indian Ocean.

Table 1.2.
Location of cores shown in Figs. 1.1 and 1.2

Core	Latitude	Longitude	Depth (m)
V28–238	01° 01′N	160° 29′E	3120
V28–239	03° 15′N	159° 11′E	3490
E49–18	46° 03′S	90° 09′E	3253
RC14–37	01° 28′N	90° 10′E	2226
V20–119	47° 57′N	168° 47′E	2739
V19–28	02° 22′S	84° 39′W	2720

Application as a formal stratigraphic scheme

The 'stages' defined originally by Emiliani (1955) enable deep-sea sediment cores of Middle Pleistocene age to be divided into rock-stratigraphic units representing intervals of time varying between 10 000 and 50 000 years. The placement of the boundaries of these stages is such that they can be correlated with an accuracy usually limited only by the degree of stratigraphic resolution possible in the presence of bioturbation in marine sediment, that is, generally to a few thousand years. We have regarded stage boundaries as being defined by their placement in core V28–238. That these stages can be used in all oceans is demonstrated by the plots in Fig. 1. 2.

Comparison with biostratigraphic methods

Traditionally marine stage boundaries are correlated by biostratigraphic means. If we accept that correlation with an accuracy of only a very few thousand years is necessary for many purposes involving Pleistocene sediments, then it might be thought unlikely that biostratigraphic techniques would be applicable. However, using the cores plotted in Fig. 1.2, Hays and Shackleton (1976) showed that the radiolarian *Stylatractus universus* Hays became extinct within the stage 12–11 boundary in all the regions studied. That is to say that particular extinction occurred essentially synchronously over a latitudinal range from 45°N to 45°S in the Pacific and Indian Oceans. This particular extinction yields such a fine degree of resolu-

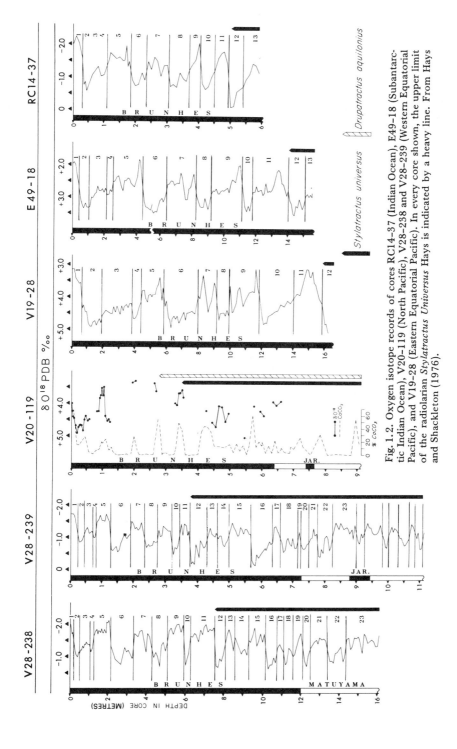

Fig. 1.2. Oxygen isotope records of cores RC14-37 (Indian Ocean), E49-18 (Subantarctic Indian Ocean), V20-119 (North Pacific), V28-238 and V28-239 (Western Equatorial Pacific), and V19-28 (Eastern Equatorial Pacific). In every core shown, the upper limit of the radiolarian *Stylatractus Universus* Hays is indicated by a heavy line. From Hays and Shackleton (1976).

tion because the organism achieved a peak in abundance shortly before extinction. Work done by Geitzenauer and others at the Lamont–Doherty Geological Observatory, as well as by Gartner (1972), suggests that the extinction of the coccolith *Pseudoemiliania lacunosa* can be utilized with equally good precision. However, work by Thompson and Saito suggests that the extinction of the Foraminiferal species *Globigerina pseudofoliata* was not a sudden world-wide event and cannot be used for fine-scale stratigraphic correlation. In any case this species had a relatively narrow geographical range (Thompson and Saito 1974 and personal communication).

It may be concluded that although we are fortunate in having at least two excellent biostratigraphic markers within sediments of Middle Pleistocene age, there is no possibility of erecting any stratigraphic structure with wide geographical applicability and appropriate refinement by biostratigraphic means alone.

Limitations on the accuracy of stratigraphic correlation

The absolute limit on the accuracy of stratigraphic correlation by means of the oxygen isotope record is imposed by the finite mixing time of the oceans. This may be illustrated as follows. As soon as glacial meltwater began to flow down the Mississippi River, about 14 000 years ago, Foraminifera in the Gulf of Mexico registered a sharp isotopic event (Kennett and Shackleton 1975). This event was presumably registered in the Atlantic surface waters almost immediately, but it was probably not recorded in Pacific deep waters for a matter of several centuries. The processes involved in the mixing of glacial meltwater into the ocean have not yet been studied, but several workers have drawn attention to the possibility that deep mixing was temporarily inhibited by the excess fresh water near the surface. Until this aspect has been investigated, an uncertainty of about 1000 years should be assumed in long-distance correlation of oxygen isotope records.

As has been mentioned, mixing by bioturbation is the more usual limiting factor. In part this is a simple matter of blurring of the details of the record by homogenization of the sediment. However, there are complications which can give rise to a systematic shift between the records of a single event in two different components of a sediment (Peng, Broecker, Kipphut, and Shackleton, in press).

Comparison of glacial stages

In principal the maximum extent by which isotopic composition in each glacial stage differs from the recent value should be a measure of the maximum ice volume accumulated during that stage, and thus should permit us to compare the extent of maximum glaciation in each. This information could assist continental workers in correlating glacial deposits with the deep-sea record.

In practice, as will be clear from the above discussion, the maximum isotopic excursion is likely to be partly obscured by bioturbation. The percentage lost will be less for the longer periods of glaciation and for those periods characterized by a high rate of sediment accumulation. Thus within cores subject to significant blurring of extremes, fluctuations in the extremes are likely. This point is illustrated in Fig. 1.3, which, using each recent value as a base-line, compares the glacial stages present in several cores of varying rates of accumulation. Cores with a low rate (e.g. V16–205, from van Donk 1976) show widely fluctuating glacial extremes. Cores with high accumulation rates (V19–29 and V19–28, Ninkovich and Shackleton 1975) show a greater recent-glacial isotopic range, and greater uniformity among the glacials. In these cores the only glacial stage showing significantly more positive values than stages 2 and 6 is stage 12 (age about 440 000 years). It has not been possible to investigate the relationship between stages 12 and 16, because no core has yet been obtained that combines a high accumulation rate with the extreme length needed to yield an excellent stage 16 sample. Thus there is an indication that more ice accumulated during stage 12 than has done since then, but there is no certainty that this was a unique event.

Correlations with the continental Pleistocene

It is not the purpose of this contribution to discuss at length the possible continental correlations with the oxygen isotope record. A good recent perspective on this subject is given by Kukla (1975) as well as by several other contributions to that same symposium volume (Butzer and Isaac 1975). However, certain pertinent points may be made here.

First, it is obvious to the writer that the continuous records of changing conditions which are being compiled for the oceans provide a very salutary yardstick against which the fragments of continental record may be placed. There is nothing in the continental records to suggest that the marine record is wrong. We know from year-by-year counts in annually laminated interglacial deposits that the time-scale of continental interglacials matches that of marine interglacial isotope stages (Shackleton and Turner 1967; Turner 1970, 1975; Müller 1974). We know from numerous radiocarbon dates that the time-scale of the last glacial matches that indicated by the oxygen isotope record. We know that the Pleistocene spans about 1·6 million years

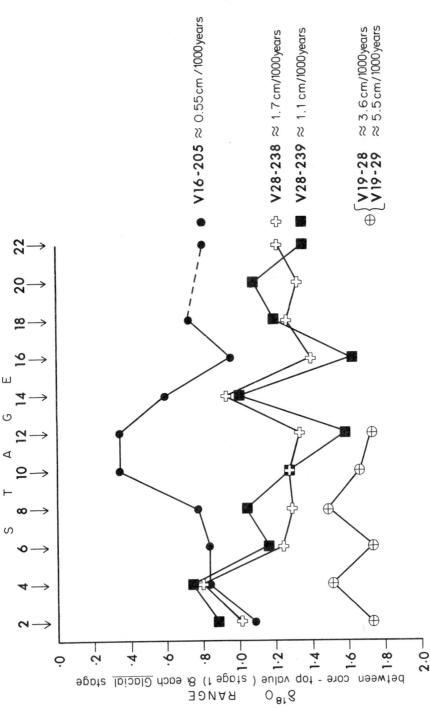

Fig. 1.3. Comparison of the amplitude of isotopic fluctuation shown in cores of various sedimentation rate. The amplitudes are expressed by showing the extent to which each glacial stage differs isotopically from the recent (core-top). The data from the two cores with highest accumulation rate suggests that the isotopic range was in reality almost the same in each successive glacial of the Middle Pleistocene. Data from van Donk (1976) for V16–205; Shackleton and Opdyke (1973) for V28–238; Shackleton and Opdyke (1976) for V28–239; Ninkovitch and Shackleton (1975) for V19–28 and V19–29.

(well reviewed by Hays and Berggren 1971). Thus it is obvious that in a country such as Britain the time represented by known and described Quaternary deposits is but a fraction of the whole of Pleistocene time.

Given these observations, it is clear that we have no internal evidence with which to assess whether the time *not* represented in known continental deposits was glacial, interglacial, or both. However, on the whole it is probably true to say that we now have, and are likely to continue to have, better continuous records for interglacials than glacials. To take an extreme case, one can conceive of obtaining a complete year-by-year description of climate and local environment throughout the entire Hoxnian Interglacial at Marks Tey (Turner 1970). Moreover, very refined correlations within the Hoxnian can be anticipated both in Britain, with other parts of Europe, and with the deep-sea record. These correlations will be accurate to 1000 years at worst, to a few decades at best. By contrast, there is little promise at present that correlations of much better than 100 000 years refinement are likely to emerge in the near future for glacial deposits following the Hoxnian. We emphasize this point because it provides a guide to those parts of the continental stratigraphic record that are likely to be useful markers for long-distance correlations (e.g. climatic deterioration at the end of interglacials) and those that are not (e.g. the base of glaciogenic deposits).

Conclusions

There are many problems, some of them outside the realms of geological science, that may be tackled using Pleistocene sediments. Many of these problems require for their solution a degree of stratigraphic refinement and chronometric accuracy that strain the resources of traditional stratigraphic techniques. Oxygen isotope methods have enormously strengthened and refined Middle Pleistocene stratigraphy.

References

Bowen, R. (1966). *Palaeotemperature analysis*. Elsevier, Amsterdam.

Butzer, K.W. and Isaac, G. Ll. (eds) (1975). *After the Australopithecines*. Mouton, The Hague.

Dansgaard, W. and Tauber, H. (1969). Glacial oxygen-18 content and Pleistocene ocean temperatures. *Science*, N.Y. **166**, 499–502.

Duplessy, J.C., Chenouard, L., and Vila, F. (1975). Weyl's theory of glaciation supported by isotopic study of Norwegian core K11. *Science*, N.Y. **188**, 1208–9.

Duplessy, J.C., Lalou, C. and Vinot, A.C. (1970). Differential isotopic fractionation in benthic foraminifera and palaeotemperatures reassessed. *Science*, N.Y. 168, 250-1.

Emiliani, C. (1954). Depth habitats of some species of pelagic foraminifera as indicated by oxygen isotope ratios. *Am. J. Sci.* 252, 149-58.

Emiliani, C. (1955). Pleistocene temperatures. *J. Geol.* 63, 538-78.

Emiliani, C. (1958). Palaeotemperature analysis of core 280 and Pleistocene correlations. *J. Geol.* 66, 264-75.

Emiliani, C. (1964). Palaeotemperature analysis of the Caribbean cores A 254–BR–C and CP-28. *Bull. geol. Soc. Am.* 75, 129-44.

Emiliani, C. (1966). Palaeotemperature analysis of Caribbean cores P 6304-8 and P 6304-9 and a generalized temperature curve for the last 425 000 years. *J. Geol.* 74, 109-26.

Epstein, S., Buchsbaum, R., Lowenstam, H.A. and Urey, H.C. (1953). Revised carbonate-water isotopic temperature scale. *Bull. geol. Soc. Am.* 64, 1315-26.

Gartner, S. (1972). Late Pleistocene calcareous nannofossils in the Caribbean and their interoceanic correlation. *Palaeogeogr. Palaeoclimatol. Palaeoecol.* 12, 169-191.

Grazzini, C. (in press). Concentrations isotopiques d'une espèce planctonique actuelle de Mer Ligure. *Palaeogeogr. Palaeoclimatol. Palaeoecol.*

Hays, J.D. and Berggren, W.A. (1971). Quaternary boundaries and correlations. In *Micropalaeontology of the oceans* (ed. B.M. Funnell and W.R. Riedel) pp. 669-691. Cambridge University Press.

Hays, J.D., Imbrie, J., and Shackleton, N.J. (1976). Variations in the earth's orbit: pacemaker of the ice ages. *Science, N.Y.* 194, 1121-32.

Hays, J.D., Lozano, J., Shackleton, N.J. and Irving, G. (1976). Reconstruction of the Atlantic Ocean and western Indian Ocean sectors of the 18 000 B.P. Antarctic Ocean. In *Investigation of Late Quaternary Paleoceanography and Paleoclimatology* (ed. R.M. Cline and J.D. Hays), *Mem. geol. Soc. Am.* 145, 337-72.

Hays, J.D. and Shackleton, N.J. (1976). The globally synchronous extinction of *Stylatractus universus. Geology.* 4, 649-52.

Hecht, A. (1976). The oxygen isotope record of Foraminifera in deep sea sediments. *Foraminifera* 2. 1-43.

Kennett, J.P. and Shackleton, N.J. (1975). Laurentide ice sheet meltwater recorded in Gulf of Mexico deep sea cores. *Science, N.Y.* 188, 147-50.

Kukla, G.J. (1975). Loess stratigraphy of Central Europe. In *After the Australopithecines*, (ed. K.W. Butzer and G. Ll. Isaac), pp. 99-188. Mouton, The Hague.

Labeyrie, (1974). Composition isotopique de l'oxygène de la silice des frustules de diatomées et spicules d'éponges. Variation de l'enrichissement isotopique avec la température. *Colloques Int. Cent. Natn. Rech. Scient.* 219, 193-201.

Lidz, B., Kehm, A. and Miller, H. (1968). Depth habitats of pelagic Foraminifera during the Pleistocene. *Nature, Lond.* 217, 245-46.

Margolis, S.V., Kroopnick, P.M., Goodney, D.E., Dudley, W.C. and Mahoney, M.E. (1975). Oxygen and carbon isotopes from calcareous nannofossils as paleoceanographic indicators. *Science, N.Y.* 189, 555-7.

Müller, H. (1974). Pollenanalytische Untersuchungen und Jahresschichtenzählungen an der holstein-zeitlichen Kieselgur von Munster-Breloh. *Geol. Jb.* A 21.

Nier, A.O. (1950). A redetermination of the relative abundance of the isotopes of carbon, nitrogen, oxygen, argon and potassium. *Phys. Rev.* 77, 789-93.

Ninkovich, D. and Shackleton, N.J. (1975). Distribution, stratigraphic position and age of ash layer 'L', in the Panama Basin region. *Earth Planet. Sci. Lett.* 27, 20–34.

Olausson, E. (1965). Evidence of climatic changes in deep-sea cores, with remarks on isotopic palaeotemperature analysis. *Prog. Oceanogr.* 3, 221–52.

Peng, T.-H., Broecker, W.S., Kipphut, G. and Shackleton, N.J. (in press). The relation of sediment mixing to the distortion of climatic records in the deep sea sediments. In *The Fate of Fossil Fuel CO_2 in the Oceans.* (eds. N.R. Anderson and A. Malahoff) Plenum, New York.

Savin, S. and Douglas, R.G. (1973). Oxygen isotope and magnesium geochemistry of recent planktonic Foraminifera from the South Pacific. *Bull. geol. Soc. Am.* 84, 2327–42.

Savin, S.M., Douglas, R.G., and Stehli, F.G. (1975). Tertiary marine paleotemperatures. *Bull. geol. Soc. Am.* 86, 1499–1510.

Shackleton, N.J. (1967). Oxygen isotope analyses and Pleistocene temperatures re-assessed. *Nature, Lond.* 215, 15–17.

Shackleton, N.J. (1968). Depth of pelagic Foraminifera and isotopic changes in Pleistocene oceans. *Nature, Lond.* 218, 79–80.

Shackleton, N.J. (1974). Attainment of isotopic equilibrium between ocean water and the benthonic foraminifera genus *Uvigerina*: isotopic changes in the ocean during the last glacial. *Colloques Int. Cent. natn. Rech. Scient.* 219, 203–10.

Shackleton, N.J. (1975). The Stratigraphic Record of Deep-Sea Cores and its implications for the assessment of Glacials, Interglacials, Stadials and Interstadials in the Mid-Pleistocene. In *After the Australopithecines,* (ed. K.W. Butzer and G.Ll. Isaac), pp. 1–24. Mouton, The Hauge.

Shackleton, N.J. (1977). The oxygen isotope stratigraphic record of the Late Pleistocene. In The changing environmental conditions in Great Britain and Ireland during the Devensian (last) cold stage. A meeting for discussion organized by G.F. Mitchell and R.G. West. *Phil. Trans. Roy. Soc.*

Shackleton, N.J. (in press). Carbon-13 in *Uvigerina*: Tropical rainforest history and the equatorial Pacific carbonate dissolution cycles. In *The fate of fossil fuel CO_2 in the oceans.* (eds. N.R. Andersen and A. Malahoff). Plenum, New York.

Shackleton, N.J. and Kennett, J.P. (1975). Paleotemperature history of the Cenozoic and the initiation of Antarctic glaciation: oxygen and carbon isotope analyses in DSDP sites 277, 279, and 281. In *Initial reports of the deep sea drilling project,* Volume 29, (ed. J.P. Kennett, R.E. Houtz, *et al.*), U.S. Government Printing Office, Washington.

Shackleton, N.J. and Opdyke, N.D. (1973). Oxygen isotope and palaeomagnetic stratigraphy of Equatorial Pacific core V28-238: oxygen isotope temperatures and ice volumes on a 10^5 year and 10^6 year scale. *Quaternary Res.* 3, 39–55.

Shackleton, N.J. and Opdyke, N.D. (1976). Oxygen isotope and palaeomagnetic stratigraphy of Equatorial Pacific core V28-239, Late Pliocene to Latest Pleistocene. In *Investigation of late Quaternary paleoceanography and paleoclimatology* (ed. R.M. Cline and J.D. Hays), *Mem. geol. Soc. Am.* 145. 449–64.

Shackleton, N.J. and Turner, C. (1967). Correlation between marine and terrestrial Pleistocene successions. *Nature, Lond.* 216, 1079–82.

Shackleton, N.J. and Vincent, E. (in prep.). Oxygen and carbon isotope studies in recent foraminifera from the Mozambique Channel region. (Also G.S.A. *Abstracts with programs*, 1975).

Shackleton, N.J., Wiseman, J.D.H. and Buckley, H.A. (1973). Non-equilibrium isotopic fractionation between seawater and planktonic foraminiferal tests. *Nature, Lond.* 242, 177–79.

Thompson, P.R. and Saito, T. (1974). Pacific Pleistocene sediments: planktonic foraminifera dissolution cycles and geochronology. *Geology* 2, 333–35.

Thorley, N. (1961). The application of geophysics to palaeoclimatology. In *Descriptive Palaeoclimatology*, (ed. A.E.M. Nairn), pp. 156–82. Interscience, New York and London.

Turner, C. (1970). The Middle Pleistocene deposits at Marks Tey, Essex. *Phil. Trans. R. Soc.* B 257, 373–440.

Turner, C. (1975). The correlation and duration of Middle Pleistocene interglacial periods in Northwest Europe. In *After the Australopithecines*, (ed. K.W. Butzer and G. Ll. Isaac), pp. 259–308. Mouton, The Hague.

Urey, H.C. (1947). The thermodynamic properties of isotopic substances. *J. Chem. Soc.* 562–81.

van Donk, J. (1976). An 0^{18} record of the Atlantic Ocean for the entire Pleistocene. In *Investigation of Late Quaternary Paleoceanography and Paleoclimatology* (ed. R.M. Cline and J.D. Hays), pp. 147–64. *Mem. geol. Soc. Am.* 145.

Vincent, E. and Shackleton, N.J. (in press). Agulhas current temperature distribution delineated by oxygen isotope analysis of foraminifera in surface sediments. *Memorial Volume to O.L. Bandy.* (ed. R. Kolpack)

2

British dating work with radioactive isotopes

F. W. SHOTTON

Abstract

The K/Ar methods are described and, although inapplicable to the British Quarternary, dates from British laboratories have contributed to East African stratigraphy. Very little work has been done in the British Isles with the uranium breakdown series and nothing with fission track dating of British samples, but the last technique is beginning to contribute significantly to Early Pleistocene studies at Lake Rudolf.

The development of radiocarbon dating laboratories in Britain is described and some of their more significant results selected. These concern particularly the ages of middle Palaeolithic, late Palaeolithic, and Mesolithic industries; the sparse remains of pre-Neolithic Man; Late-Devensian and Flandrian stratigraphy following the last glacial maximum; and preglacial Devensian stratigraphy and the dating of climatic fluctuations back to the Chelford (Brørup) Interstadial.

Brief mention is made of the use of ^{14}C (and ^{13}C) and of tritium in hydrogeological work.

Introduction

Because unstable isotopes break down at a rate which is independent of physical conditions, their importance to geology, and to the Quaternary in particular, is that this decay may be used to measure the passage of time. This is radiometric dating and on the assumption that readers are already familiar with the principles governing its application, it may be simply stated that there are three isotopes, or isotope series, which may be used for the Quaternary. These are:

(i) potassium 40 with its production of argon 40 (K/Ar method);
(ii) the two uranium breakdown series leading to lead as end product;
(iii) carbon 14.

Fission track dating may also be included as a method based on the instability of ^{238}U.

This chapter will examine the extent to which these unstable isotopes have been used in British laboratories or to solve British problems.

The $^{40}K/^{40}Ar$ and $^{40}Ar/^{39}Ar$ methods

^{40}K breaks down to ^{40}Ar and ^{40}Ca, and occurs as a small amount in proportion to the ^{39}K in newly crystallized minerals such as potash feldspars, muscovite and other micas, hornblende or glauconite forming in marine sediments. If the radiogenic argon is totally retained in the sample, it may be liberated and measured; the ratio $^{40}Ar/^{40}K$ gives a measure of the time over which breakdown has occurred. If ^{39}K is irradiated with fast neutrons it produces ^{39}Ar, and the ratio of $^{40}Ar/^{39}Ar$ provides another basis for age determination. This may be pursued by the technique of total degassing or by the more complicated process of step heating developed by Miller (Miller 1972, Fitch 1972).

Most of the dating of Quaternary and late Caenozoic rocks has been done in the United States but British scientists have determined a considerable number, notably on several volcanic horizons at the Artifact Site of Lake Rudolf (Fitch and Miller 1970, 1976). Perhaps their most important contribution, however, is their argument that the step heating $^{40}AR/^{39}Ar$ method of dating allows reliable samples to be distinguished from those where some loss of radiogenic argon has taken place (Fitch *et al.* 1976a). On this contention they dated the artifact-containing KBS Tuff of Lake Rudolf at 2·61 ± 0·26 million years, which is 0·4–0·8 million years older than some of the figures obtained by conventional degassing methods. In an even more recent publication (Fitch, Hooker, and Miller 1976b) the most acceptable date for the same horizon is revised to 2·42 ± 0·01 million years.

Because the half-life of ^{40}K is long, the method is less subject to a large proportionate error the older the sample is. It is important in the early Pleistocene, but only in exceptional cases can the method measure ages which overlap the radiocarbon field and normally the estimation becomes untrustworthy at ages younger than about 250 000 years.

The argon methods can only be used with lavas, tuffs or, less reliably, glauconitic marine sediments. Since there are no volcanic rocks and virtually no glauconitic sediments in the British Quaternary, the methods have not been applied to the United Kingdom.

The transient isotopes in the breakdown chains of ^{238}U and ^{235}U to Pb

The two uranium isotopes exist in small but virtually constant amounts as compounds in sea water. As ^{238}U breaks down towards ^{206}Pb and ^{235}U to ^{207}Pb, certain transients such as thorium–230, protoactinium–231, and radium–226 appear and decay with half-lives which are commensurate with Quaternary time. There are several ways in which the proportions of these transients may be used for age determination (Broecker 1965), but as the methods are mainly applicable to deep-sea sediments, they are not relevant to the British Pleistocene. Another method applicable to calcareous skeletons of organisms depends on the incorporation during growth of some uranium, which then starts to break down, producing ^{230}Th. This is also unstable and eventually an equilibrium between thorium production and decay is reached. Until then, the changing ratio $U/^{230}Th$ should be a reflection of age; but there are several possible sources of error (Shotton 1967). Although the method seems satisfactory when used with corals (which incorporate an exceptional amount of U), it seems to be unreliable when applied to molluscs. As a consequence, little work on these lines has been done in this country. Reference should be made, however, to a pioneer study by Szabo and Collins (1975) on the uranium derivatives in bones in certain British sites. For Clacton (Essex), equated with part of the Hoxnian, they gave a figure of 245 000 ± 25 000 and for a bone from Swanscombe only a few centimetres below the horizon of the well-known skull, >272 000. A deposit at Brundon in Essex, usually regarded as Ipswichian, gave a figure of 174 000 ± 30 000 and at Stutton (Suffolk) from brickearth 0·5 m above Ipswichian deposits, 125 000 ± 20 000 resulted. These results have large probability errors but they at least give some idea of the dates of these two interglacials and suggest an interval of the order of 100 000 years between them.

Fission Track Dating

This depends upon the slow but steady rate at which atoms of ^{238}U divide themselves explosively, developing in the containing minerals atom-disorientated tracks which may be revealed by etching (Fleischer and Hart 1972). The method is only applicable to minerals and glasses in volcanic rocks within a Quaternary succession and the absence of such in Britain has inhibited work by British scientists. Nevertheless, Fitch's laboratory is developing the method and has obtained consistent results using zircon lumps from the pumice of the KBS Tuff of the Koobi Fora formation of East Rudolf (Hurford *et al.*1976). The figure they obtain of 2·42 million years is strikingly in harmony with that by the argon step-heating method mentioned earlier and it suggests that previous K/Ar determinations of about 1·8 million years are under-estimates.

Carbon-14 (radiocarbon)

It is unncessary here to explain why the radioactive isotope of carbon is, and has been in the past, present as a minute part of the CO_2 of the atmosphere and of that dissolved in the oceans. Living organisms incorporate it into their cells, either directly by photosynthesis or less directly via food chains, in virtually the same proportion to stable carbon as exists in the atmosphere. Only after death of the organism does the ^{14}C start on its irreversible course of disintegration. It has a half-life now accepted as 5730 years, but so many dates have been published on the basis of an earlier determination of 5570 years that it is still the custom to continue with this practice. Dates are thus 3% smaller than they should be and we now also know that they differ, at least over the last 8000 years, from the figures of dendrochronology.

It is not necessary to explain how the pitfalls of radiocarbon dating—the Suess effect, isotopic replacement, hard water error, and the contamination of samples either by ancient inert carbon or modern organic material—are avoided or corrected. What is essential to a proper understanding of the method's technology, however, is the appreciation of the proportions of the three different carbon isotopes in atmospheric CO_2. ^{12}C (stable) makes up nearly 99%, ^{13}C (stable) just over 1%, and ^{14}C one part in 10^{12}. Even in living organic matter, therefore, this minute proportion of ^{14}C is not measurable, as are ^{12}C and ^{13}C, by the mass spectrometer. The only method that can be used is the counting of the electrons which are emitted from a sample as the ^{14}C atoms break down and comparison of their frequency with that produced from an equal-sized sample of modern material. Basically there are two ways of doing this. Gas prepared from the sample can be enclosed in a container down the centre of which runs a wire carrying a high prescribed positive charge. The determination of the value of this (the 'plateau') is slightly complicated.

Discharged electrons are attracted to and neutralized by this positive charge, and they may be counted by electronic means. This is *gas proportional counting*. Alternatively the original sample may be converted into a suitable liquid hydrocarbon (now almost inevitably benzene), laced with a suitable non-radioactive scintillator which will flash when an electron is released; these flashes can be counted photoelectrically. This is *liquid scintillation counting*.

Both processes are much more complicated than is suggested by the basic description here. One complication which is common to both is that part of the cosmic radiation which permeates the space of the laboratory can, whatever steps are taken to minimize it, get through to the counting chamber and there be registered in addition to the natural activity of the sample. So inevitably an equal sample prepared from ancient material devoid of ^{14}C (for example, Carboniferous anthracite) has to be counted in addition to the material under assay and it is the difference between these two results which is compared with the activity of modern material, previously determined under the same conditions. Since atomic disintegration is a random process, every determination must be accompanied by a figure which expresses the probability of deviation from a long-term average. This is computed by a simple statistical process but it must always be remembered that when the counting figures are turned into ages, the statement of an age of 2000 ± 80 does not mean that the sample age lies between 1920 and 2080 years, but only that there is a 2 to 1 chance that it does so.

Proportional gas counting was perfected before scintillation counting, which tends to be preferred in the more recent installations. The size of a gas counter (up to 6 litres) is so large compared with the few millilitres of a benzene-filled vial, that much more elaborate means of shielding from cosmic radiation have to be used. Typically the counter will be surrounded by a multi-anode guard ring or a battery of geiger tubes designed to filter off from the counting process most of the β-particles (electrons) which have got that far, and all this will be within a 'castle' of lead or steel, possibly with a shell of liquid mercury and another of borated paraffin wax to absorb neutrons. The gas in the counter may be the purified CO_2 from combustion of the sample or it may be converted to another compound. In British laboratories the choice lies between CO_2 and CH_4, the latter being less absorptive of its own β-particles and hence capable of being measured in larger amounts. A limiting factor in scintillation counting is that benzene must be produced by a process of the type

$$\text{Sample} \xrightarrow{\text{combustion}} CO_2 \xrightarrow{} \text{a carbide} \xrightarrow{\text{+ water}} C_2H_2 \xrightarrow{\text{polymerization}} C_6H_6 .$$

As all water now contains tritium, which is radioactive, so also will the benzene. By reducing the counting efficiency to about 65%, the weaker tritium disintegrations are not registered, but this lowers the ultimate dating limit of the apparatus.

Changes in barometric pressure cause small variations in the 'background count' and because in gas-proportional counting the sample and the inert control are measured consecutively, a small barometric pressure correction has to be applied. This is not necessary in the scintillation method where sample and control are assayed more or less concurrently.

Brief details of British radiocarbon laboratories

British Museum (BM)

Originally gas proportional with C_2H_2. Because of tritium trouble, changed in 1964 to CO_2. Now converted to benzene liquid scintillation. First list 1959.

Cambridge: Sub-department of Quaternary Research (Q)

Gas proportional with CO_2. In 1970 changed to scintillation guards but retained gas counting. First list 1959.

Dublin (D)

First list in 1961 based on a short-term experiment with liquid scintillation using methyl alcohol. Re-started in 1974 using benzene.

National Physical Laboratory (NPL)

Gas proportional with CO_2 and standard Geiger anticoincidence guard ring. First list 1963, final list 1969. Discontinued.

Birmingham University Geology Department (Birm)

Gas proportional with CH_4, 1·2 and 6 litre counters working to 3 atmospheres with Geiger anticoincidence guards. Later a third counter, Oeschger type, added. First list 1967.

Belfast University (UB)

Gas proportional with CH_4. 2·2 litre counter working to 5 atmospheres. Multi-anode proportional guard counter, methane filled. First list 1970.

Glasgow University (GU)

A double system of two 0·5 litre counters filled to 5 atmospheres with CH_4 and surrounded by a methane-filled guard. Since converted to tritium counting. A 2·5 litre gas counter using CO_2 at 1 atmosphere with a Geiger gas-flow guard. A conventional benzene liquid scintillation counter added in 1974. First list 1969.

Scottish Universities Research and Reactor Centre (SRR)
Originally two benzene liquid scintillation counters, a third now added plus a small gas proportional counter for experimental purposes. First list 1973.

A.E.R.E. Harwell (Ha)
Originally a laboratory for counting tritium. Two benzene liquid scintillation counters added for radiocarbon work. First list 1974.

The range of activity of British laboratories
Fig. 2.1 shows, in histogram form, the number of determinations made in most of the British laboratories. In explanation, it should be stated that:

> (a) The records are taken from all lists published in *Radiocarbon* up to April 1976, with additional knowledge of what was in press for Birmingham and Scottish Universities Research and Reactor Centre.
> (b) Although the dates were grouped at 3000 year intervals down to 15 000 and in 5000 year groups after that, the columns show an averaged frequency over each 1000 years, i.e. if there were 185 dates between 3000 and 6000 years, the column will rise to 61·7 but if there were 26 dates between 25 000 and 30 000, it will stand at 5·2.
> (c) The histograms were plotted on a logarithmic scale in order to make more clear the infrequent but very significant dates at the older end of the range.

Dublin has been omitted because of the limited time over which it has operated and Harwell because it has only published one list—which does it less than justice since its second list will be considerable. Also omitted is Glasgow which has largely concentrated on variations in the modern atmosphere affected by atomic bomb explosions and on the composition of modern vegetation and water. It has now embarked on a considerable programme of dating connected with 'floating' dendrochronologies, but results are not published.

These histograms not only show the size of a laboratory's activity, but also its range. They also give some idea of the oldest limits of positive dating within the laboratory's capacity.

All laboratories have a concentration of dates in the first 6000 years, but the most polarized is Belfast which has nothing older than 15 000. This reflects its policy of concentration on the archaeology of Northern Ireland and the development of the postglacial flora. The British Museum has 96% of its dates younger than 10 000 and this again reflects its prime purpose, which is to be a collaborator with archaeology. The weighting of Cambridge's activity towards

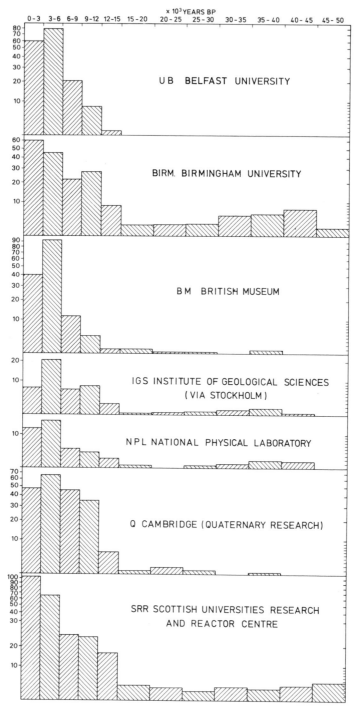

Fig. 2.1. Histograms showing the distribution of determined radiocarbon ages at most of the British laboratories.

the zonation of the postglacial Late Devensian and the Flandrian is also clear.

For some years before the Scottish Universities Research and Reactor Centre became operational, the Institute of Geological Sciences submitted samples to Stockholm which were published in *Radiocarbon*. These records have also been summarized in Fig. 2.1.

Research on the Devensian earlier than the main glacial advance demands dates going back as far as possible. Birmingham University's large methane gas counter was designed with this in mind. It allows the quoted standard deviation of the activity of an old sample to be as low as 0·06% of modern activity, which permits the quoting of dates to a limit of 49 000 years. The SRR laboratory can achieve the same limit with liquid scintillation counting and NPL, whilst it was working, could reach 45 000 years. These possibilities are apparent in Fig. 2.1, although many of the 'dates' in the 45 000–50 000 bracket are 'greater than' figures. No British laboratory has emulated the technique of diffusion enrichment practised by Gröningen (GrN), with the possibility of dating to about 70 000 years.

Some of the more important results
1. Palaeolithic and Mesolithic industries
Stone implements can only be dated when they are included in contemporaneous deposits containing bone, wood or charcoal or, less reliably, shell. In the lists which follow, dates are simplified by quoting them without standard deviation and the principal determining laboratories are indicated by their index letters.

(1a) Middle Palaeolithic
Mousterian of Acheulian tradition, Coygau Cave, Dyfed, Wales. 38 684 (BM).
(1b) Upper Palaeolithic
Early Upper Palaeolithic leaf points. Kent's Cavern, Devon. 28 160, 28 720 (GrN).
Possible Proto-Solutrean leaf points. Robin Hood's Cave, Derbyshire. 28 500 (BM).
Blade of Magdalenian type. Caldey Island, S. Wales. 22 350 (Birm).
Creswellian. Creswellian Point Industry. One early date from Sun Hole, Somerset, 12 378. Five from Robin Hood's Cave, Derbyshire, and Anston Stones Cave, S.Yorkshire, between 10 590 and 9750 (all BM).
(1c) Mesolithic
Maglemosian. From Thatcham and Greenham, Berkshire and Starr Carr, Yorkshire, 9 dates between 10 365 and 8779 (mainly Q, one C (Chicago)). Chronological overlap with the Creswellian seems to be indicated.
(1d) Scottish raised beach shell middens
The shell mounds on the Isle of Oronsay in Western Scotland lie on a low raised beach whose shells have been dated between 7420 and 7020 (Birm). A Birm. date for 'occasional occupation' of 6910 and for regular occupation

four dates between 6010 and 5150 (Birm) and also 5755 and 5015 from GX. (Geochron Laboratories Inc., Cambridge, Mass.)

2. Early Man

Pre-Neolithic human skeletal material is not common in Britain and the most venerable Swanscombe skull is far beyond the reach of radiocarbon. The remains of a few humans have, however, been dated on their collagen.

> The 'red lady of Paviland' (actually a male), S. Wales. 18 460 (BM).
> 'Cheddar Man', Somerset. Associated with Creswellian or Mesolithic industry. 9080 (BM).
> Bones in Aveline's Hole, Somerset. Associated with Creswellian or mesolithic industry. 9114 (BM).

3. Dates from the Flandrian and postglacial late Devensian

There are many hundreds of determinations within this range of time and it is difficult to select those of especial significance. One of the more puzzling dates comes from a moss layer at Glen Ballyre, Isle of Man, which lies above late-Devensian till. It was dated four times on two separate collections (Birm 213, 270a, b, c) all figures lying between 18 900 ± 330 and 18 400 ± 500, and this is so unexpectedly old that the possibility of the moss, *Drepanocladus revolvens*, photosynthesizing under hard water must be conceded. Shotton (1972), measuring dates in an Allerød succession at Norre Lyngby, Denmark, showed that there could be a hard water error of the order of 1700 years between carbonaceous muds (thought to be largely subaqueous algal) and fragments of leaf and twig.

Cambridge (see, for example Godwin 1960) undertook a dating programme at Scaleby Moss and Red Moss, Lancashire and Tregaron Bog, Wales, to fix the limits of the British Flandrian pollen zones. Belfast did a similar task in Northern Ireland. There was no obvious evidence of diachronism and the following figures, based on about 33 determinations, must be close to the truth:

VIIb/VIIa, 5100 BP; VIIa/VI, 7120 BP; VI/V, 8700 BP; V/IV, 9750 BP; IV/III, 10 200 BP.

In addition, the III/II boundary is close to 10 800 and II/I to 12 000.

4. Preglacial Devensian dates

Although less than 300 determinations (8·5%) give dates older than 20 000 years, and this total includes a number of 'infinites', they are of disproportionate importance in the elucidation of Devensian stratigraphy and it is difficult to select only a few for special mention.

Two of the most unexpected (in this case newer than 20 000) were the dates on the moss bed of Dimlington (south Yorkshire), 18 500 (I–3372) and 18 240 (Birm–108) proving the late-Devensian tills to be younger than previously supposed. A somewhat similar situation occurs in North Wales at Tremerchion (Rowlands 1971). Cave deposits with a cold climate mammal fauna and artifacts conventionally regarded as Aurignacian, are sealed by late-Devensian till. A date of 18 000 ± 1200 (Birm–146) on a bone of mammoth suggests a late date for the ice advance, much as at Dimlington. At the same time it arouses doubts about the 'Aurignacian' artifacts. Either they belong to a later period of the Upper Palaeolithic, or the cave deposits range through a considerable period of time and one of the latest of the bones was assayed. Alternatively, of course, the bone which came from an old collection might have been contaminated, even though the standards of cleansing were meticulously applied.

There have been many dates on river terrace material, between 38 000 and 26 000, proving the approximate contemporaneity of No. 2 Terrace of the Warwickshire Avon and the Carrant Brook, the Low Terrace of the Tame, No. 1 Terrace of the Nene, and the Upper Floodplain Terrace of the upper and middle Thames. Associated with these terraces is a cold tundra fauna which includes the mammals *Mammuthus primigenius, Coelodonta antiquitatis, Ovibos moschatus, Rangifer tarandus, Bison*, and various lemmings.

The interpretation of the type Devensian section at Four Ashes (Staffordshire) (Morgan, A.V. 1973; Morgan, Anne 1973) and the recognition therein of the two interstadials of Upton Warren (in its limited sense) and Chelford was greatly helped by the radiocarbon dates. The same short-lived interstadial of Upton Warren, with its warm peak around 43 000 BP, has been demonstrated by collaboration between workers on the insect fauna and in the dating laboratories of Gröningen and Birmingham, at Upton Warren (Worcestershire), Tattershall (Lincolnshire), Earith (Huntingdonshire), and Isleworth (Middlesex).

The two most important dates in Ireland are at Derryvree (Fermanagh) where silts dated at 30 500 (Birm) lay between two tills, and at Castlepook Cave, Co. Cork, where a mammoth bone gave an age of 33 500 (D–122). So far this is the only dating of mammoth and its extensive associated vertebrate fauna in Ireland.

In the extreme north of Shetland, a peat lying between two tills at Fugla Ness has been given a date of 40 100 (SRR).

Finally, mention must be made of the only British 'enriched' date, done by Gröningen on the interstadial deposits of Chelford (Cheshire). At 60 800 ± 1500 (GrN 1480) it strengthens the botanical correlation of Chelford with the Brørup Interstadial of Holland and Denmark.

Work on water using carbon and tritium

Hydrogeologists use the apparent radiocarbon age of carbonate dissolved in water, and also its content of stable ^{13}C, as indicators of the importance or otherwise of contemporary infiltration into an aquifer. Although this can differentiate a recharging aquifer from one with 'fossil' water, the method bears only marginally on Quaternary studies. The accession, after the first explosions of atomic bombs, of abnormal amounts of tritium into the atmosphere and so into rain, has enabled the element to be used as a tracer in groundwater and as a means of determining the time required for rainfall to reach the water table. It is also a clear indicator of whether an aquifer is being recharged or not, since the groundwater contains tritium in the first case but negligible amounts in the second. The decay rate of tritium is shown by its half-life of 12·26 years and the isotope will soon become so unimportant that it will be useless for this sort of work, unless it is regenerated by new atomic bomb explosions. There are few scientists who would welcome this.

Acknowledgements

I am grateful to the directors of various laboratories and in particular to H. Barker, M.S. Baxter, R. Burleigh, F.J. Fitch, D.D. Harkness and R.L. Otlet, for giving me up-to-date information about their installations and the progress of their work.

References

Broecker, W.S. (1965). Isotope geochemistry and the Pleistocene climatic record. Lamont Geol. Observ. Contribution 792. In *The Quaternary of the United States* (ed. H.E. Wright, and O.G. Frey), pp. 737–35.

Fitch, J.F. (1972). Selection of suitable material for dating and the assessment of geological error in potassium-argon age determination. In *Calibration of hominoid evolution* (ed. W.W. Bishop, and J.A. Miller), pp. 77–91. Scottish Academic Press and University of Toronto Press.

Fitch, J.F. and Miller, J.A. (1970). Radioisotopic age determination of Lake Rudolf artefact site. *Nature, Lond.* 226, 226–8.

Fitch, J.F. and Miller, J.A. (1976). Conventional Potassium–Argon and Argon 40–Argon 39 dating of volcanic rocks from East Rudolf. In *Earliest man and environments in the Lake Rudolf Basin.* (ed. Y. Coppens *et al*), 123–147. University Press, Chicago.

Fitch, J.F., Hooker, P.J. and Miller, J.A. (1976a). Geochronological problems and radio-isotopic dating in the Gregory Rift Valley. In *Geological background to fossil man.* (ed. W.W. Bishop), *Geol. Soc. Lond.* Scottish Academic Press.

Fitch, F.J. Hooker, P.J. and Miller, J.A. (1976b). Argon-40/argon-39 dating of the KBS Tuff in Koobi Fora Formation East Rudolf, Kenya. *Nature, Lond.* 263, 740–44.

Fleischer, R.L. and Hart, H.R. (1972). Fission track dating techniques and problems. In *Calibration of hominoid evolution*. (ed. W.W. Bishop and J.A. Miller) pp. 138-70. Scottish Academic Press and University of Toronto Press.

Godwin, H. (1960). Radiocarbon dating and Quaternary history in Britain. *Proc. R. Soc.* B 153, 287-320.

Hurford, A.J., Gleadow, A.J.W. and Naeser, C.W. (1976). Fission track dating of pumice from the KBS Tuff, East Rudolf, Kenya. *Nature, Lond.* 263, 738-40.

Miller, J.A. (1972). Dating Pliocene and Pleistocene strata using the potassium— argon and argon-40/argon-39 methods. In *Calibration of hominoid evolution* (ed. W.W. Bishop and J.A. Miller), pp. 63-76. Scottish Academic Press and University of Chicago Press.

Morgan, A. (1973). Late Pleistocene environmental changes indicated by fossil insect faunas of the English Midlands. *Boreas* 2, 173-212.

Morgan, A.V. (1973). The Pleistocene geology of the area north and west of Wolverhampton, Staffordshire, England. *Phil. Trans. R. Soc.* B 265, 233-97.

Rowlands, B.M. (1971). Radiocarbon evidence of the age of an Irish Sea Glaciation in the Vale of Clwyd. *Nature, Lond.* 230, 8-10.

Shotton, F.W. (1967). The problems and contributions of methods of absolute dating within the Pleistocene period. *Jl. geol. Soc. Lond.* 122, 357-83.

Shotton, F.W. (1972). An example of hard water error in radiocarbon dating of vegetable matter. *Nature, Lond.* 240, 460-1.

Szabo, B.J. and Collins, D. (1975). Ages of fossil bones from British interglacial sites. *Nature, Lond.* 254, 680-82.

3

British Quaternary non-marine Mollusca: a brief review

M. P. KERNEY

Abstract

Terrestrial and freshwater molluscs are valuable for the interpretation of ecology and also of climate. Certain brackish-water species (e.g. of *Hydrobia*) are useful indicators of adjacent marine transgressions. As with plants, the succession of species in general follows a similar pattern in all the interglacials, but there are differences of detail. Molluscs are not a rapidly evolving group but there are changes of fauna with time, attributable more to ability to migrate to or from the continent than to evolutionary change. Detailed consideration is given to the faunal assemblages of the cold periods, particularly the Devensian, and to the various interglacial stages from pre-Cromerian to Flandrian. Limited criteria for stratigraphical separation are listed.

Introduction

Land and freshwater shells are among the commonest of Pleistocene fossils and have attracted scientific attention in Britain at least since the time of Morton (1706). A.S. Kennard (1871–1948) published over 250 papers on the subject. Inevitably the aim of much of this earlier work was taxonomic rather than environmental. Furthermore, Kennard himself perhaps over-estimated the importance of non-marine molluscs as tools for direct dating in a classic geological sense—a task for which they are inherently rather badly fitted. For these reasons, at the period when palynology began to yield important and more easily interpretable results, the Mollusca were for a while neglected.

A fresh approach may be traced to the later 1950s. Sparks (1957a) showed how a detailed quantitative analysis of faunal changes through an interglacial deposit could provide valuable ecological and climatic evidence. Many similar papers have since appeared and the more important are listed at the end of this chapter.

General considerations

The basis of any analysis must depend on an understanding of the stratigraphy, and on precise sampling. Techniques of sampling, extraction, and counting are described by Sparks (1961, 1964c) and Evans (1972). Most non-marine Mollusca are small, and the distortions which arise if only the large or unbroken snails are extracted are strikingly illustrated by Sparks (1961, Fig. 1).

The living fauna of the western palaearctic region is fortunately rather fully known. Identification of fragmentary fossil material does nevertheless present problems not generally encountered by the biologist, demanding a knowledge of juvenile stages or shell microsculpture, subjects that are poorly covered in existing literature.

Nomenclature is now reasonably stable. For the gastropods the system of Wenz and Zilch (1938–60) is now broadly followed in most European countries, and revised check-lists have recently been published for Britain (Waldén 1976, Kerney 1976b).

The first attempt at expressing snail counts graphically was by Burchell, who used sector-diagrams to show the relative proportions of four selected species in Neolithic and Bronze age deposits in Kent (Burchell and Piggott 1939). Sparks (1957a, Fig. 1) published the first true molluscan diagram, showing the vertical percentage changes within the total fauna through the sediments of the Ipswichian type site. This diagram is of the 'saw-tooth' form often employed by palynologists; later it became customary to use histograms, showing the thickness of each sample precisely.

Occasionally it may be useful to express fluctuations in terms of absolute frequencies per unit of sediment. This has proved illuminating when dealing with periglacial slope deposits (Kerney 1963, 1971*b*).

The way in which shells accumulate must first be considered. Fossil assemblages rarely represent living communities. An open river deposit may contain sweepings from a wide spectrum of freshwater and terrestrial habitats indiscriminately brought together, whereas a sediment deposited in a quiet lake may yield only a very limited number of species living in the water above the spot. A comparison between contemporary river and lake deposits of Hoxnian Zone II date shows this clearly (Kerney 1971*a*; Sparks *et al.* 1969). Similarly, within slope deposits strong changes may be noted between the ecologically mixed assemblages in colluvial layers and the more homogeneous assemblages in fossil soils.

Ecological reconstruction must begin with knowledge of the modern fauna. Sometimes the behaviour of a single genus may demonstrate an environmental change very clearly, as in the sudden incursion of brackish-water *Hydrobia* at the level of marine transgression within the Ipswichian interglacial deposits at Selsey (West and Sparks 1960, Fig. 7). Grouping species into ecological categories may be very helpful in illuminating broad changes in the environment, e.g. swamp as against open water, or woodland as against grassland. Sparks (1961) has found a fourfold categorization of all freshwater species into a 'slum group', a 'catholic group', a 'ditch group', and a 'moving water group' consistently helpful. Groupings of this kind are additionally often employed to prevent the diagrams becoming too congested and losing meaning to the eye. However, it must be accepted that all such generalizations are somewhat subjective, since the known environmental ranges of many molluscs are not sharply definable and may be remarkably wide. Ecological tolerance is also a function of climate, many snails becoming increasingly fussy as the limits of geographical range are approached. Knowledge of ecology remains imperfect above a simple level.

Paradoxically, it is sometimes possible from quantitative studies of fossil communities to form useful hypotheses about ecology which have not yet been tested by appropriate research on living faunas. A good example is the discussion by Evans (1972, p. 153) of the precise requirements of the British species of the common grassland genus *Vallonia*.

Climatic interpretation is usually based on knowledge of present-day geographical range. For example, the southward retreat of the Meditterranean–Atlantic snail *Pomatias elegans* to its present north-easterly limits in Europe can plausibly be linked with a drop in winter temperatures since the Flandrian optimum (Kerney 1968). Sparks (1961) has shown how the changing proportions of groupings based on present distributions can reveal clear trends of either improving or deteriorating climate within Ipswichian sequences. In addition, a useful guide is provided by the relatively detailed picture now emerging for the changes which have occurred since the Late Devensian. Nevertheless, models based on Flandrian or present-day patterns often break down if applied too rigidly to earlier periods.

Faunas of the British Quaternary

Evolution has been rather slow. If we consider the freshwater species alone (knowledge of the terrestrial species is much more fragmentary) we find that out of 57 indigenous Flandrian species, 41 (72%) are known from pre-Anglian deposits. Conversely, only about 9 species found in pre-Devensian deposits are extinct (marked with a dagger (†) in the account below). The changes in the faunas must therefore be explained largely in environmental terms, or in terms of different patterns of migration consequent on climatic change.

Glacial faunas

These are usually composed of a small number of species (commonly 10 or less), but a few may occur abundantly. Holarctic or palaearctic elements predominate. The terrestrial Mollusca are of open-ground character.

Glacial faunas are very poorly known for stages prior to the Devensian, which will therefore be considered first.

Devensian. It is not clear whether any Mollusca survived in England through the period of maximum cold of the Late Devensian some time after 26 000 BP; no faunas can be assigned with certainty to a time between this date and *c*. 14 000 BP (late glacial).

Middle Devensian faunas are known from a number of sites, dated by radiocarbon between *c*. 42 000 BP (Upton Warren; Coope, Shotton, and Strachan 1961) and 27 650 ± 250 BP (Beckford; Briggs, Coope, and Gilbertson 1975). Twenty-three species are recorded, although only nine occur with any frequency (*Lymnaea truncatula, L. peregra, Anisus leucostoma, Gyraulus laevis, Succinea oblonga, Oxyloma pfeifferi, Columella columella, Pupilla muscorum, Deroceras* sp.). There is a mixture of catholic palaearctic and of arctic–alpine elements, the latter now extinct in Britain (*Columella columella, Vertigo genesii genesii, Pisidium obtusale lapponicum, P. vincentianum*). Among the freshwater species the common presence of

Gyraulus laevis is often distinctive; and among the land Mollusca *Pupilla muscorum*, a snail of poorly vegetated grassy places (e.g., screes and calcareous dunes) may occur in very large numbers, giving a characteristically 'glacial' facies to the assemblages.

Slightly milder episodes of 'interstadial' character are possibly reflected at certain Middle Devensian sites by high proportions of *Trichia hispida*, a snail whose northern limit in Europe may be climatically determined (Kerney 1971*b*).

The way in which these impoverished faunas developed during the Early Devensian (pre-*c.* 50 000 BP) is not yet clear. Species of relatively southern range survived in reduced numbers from the Ipswichian for a considerable while. At Wretton (West *et al.* 1974) and Isleworth (Kerney 1959*a*) fairly rich freshwater faunas with temperate elements (*Bithynia, Anisus vortex, Pisidium moitessierianum*) are associated with purely open ground *Pupilla–Trichia* assemblages. And at Pitstone, Buckinghamshire, a fossil soil complex within chalky solifluxion, possibly assignable to the Chelford Interstadial, contains a remarkable terrestrial fauna dominated by a mixture of *Pupilla, Trichia*, and *Azeca goodalli*, coupled with small numbers of the thermophile *Discus rotundatus*, but with no northern elements (Evans and Kerney, in preparation).

Turning to the end of the Devensian, late-glacial faunas are relatively well-known and comprise about 40 species. The freshwater assemblages remain restricted, and in lake sediments characteristically show a monotonous facies with abundant *Valvata cristata, V. piscinalis, Lymnaea peregra, Gyraulus laevis*, and *Armiger crista*. Terrestrial faunas are best known from calcareous slope deposits in south-east England and can there be related to a lithostratigraphy in which a buried soil of the Allerød Interstadial forms an important marker (Kerney 1963). The assemblages are considerably richer than those of the Middle Devensian and offer a curious mixture of biogeographical elements: (*a*) climatically tolerant species of wide range (e.g., *Pupilla muscorum, Punctum pygmaeum, Nesovitrea hammonis, Euconulus fulvus*), (*b*) arctic–alpine species (e.g., *Columella columella*), (*c*) species of limited range in northern Europe (e.g., *Trichia hispida*); and (*d*) west European species absent from Scandinavia (e.g., *Abida secale, Helicella itala, Trochoidea geyeri*). Elements (*c*) and (*d*) appear strongly during the Allerød Interstadial, and persist throughout the Younger Dryas. As with the flora of the late-glacial, these assemblages have no modern analogue.

Pre-Devensian. Glacial faunas of probably Wolstonian age are described from solifluxion deposits at Thriplow (Sparks 1957*b*) and from river deposits at Long Hanborough (Briggs and Gilbertson 1973). Both are very similar to those from the Middle Devensian, the terrestrial assemblages characterized by *Pupilla* and *Columella columella*. The occurrence of the marsh snail *Catinella arenaria* is a further link with certain Middle and Late Devensian deposits: the species is nowhere known with certainty from the interglacials, and today is virtually restricted to sand-dune habitats along the Atlantic coast of western Europe.

An early Wolstonian assemblage from Thames deposits at Swanscombe (Kerney 1971*a*) suggests the survival from the previous interglacial of a number of relatively thermally demanding freshwater species, coupled with a *Trichia–Pupilla* fauna of open-ground character (cp., the Early Devensian).

Anglian late-glacial faunas of lacustrine facies are described from Hoxne (Sparks *in* West 1956) and from Hatfield (Sparks *et al.* 1969). They are dominated by *Gyraulus laevis*, *Armiger crista*, and *Valvata piscinalis*, and offer a clear parallelism with many Devensian late-glacial lake sediments.

Interglacial faunas

These are characterized by being rich in species (sometimes 70 or more). A significant proportion may belong to biogeographical groups with restricted northerly ranges, and some may no longer live within the British Isles (e.g., *Corbicula fluminalis*). In pre-Flandrian deposits extinct species may occur. In the middle parts of interglacials a number of woodland snails may be found, though their frequency will obviously depend on the nature of the deposits. The least informative assemblages are usually those from isolated lake basins, often composed only of a small number of both ecologically and climatically tolerant freshwater species, from which no very useful picture of regional changes can be deduced (e.g., at the Hoxnian type site, and in many Flandrian lake marls).

Taking freshwater faunas first, knowledge is most complete for the Ipswichian (Sparks 1964*b*). Ip I assemblages are restricted, and the late-glacial facies *Gyraulus laevis–Armiger crista* may persist (Sparks and West 1968). Maximum species diversity appears to have been reached in Ip IIb, and the highest proportions of species with restricted northward ranges in Ip IIb and III. Characteristic warm indicators are *Belgrandia marginata*, *Segmentina nitida*, *Potomida littoralis*, *Corbicula fluminalis*, †*Pisidium clessini*, *P. supinum*, and *P. moitessierianum*. But these temperate elements may persist in diminishing numbers throughout Ip IV, for example *Corbicula* at Stutton (Sparks and West 1964).

Knowledge of Hoxnian freshwater faunas is less complete, but differences emerge. *Anisus vorticulus, Gyraulus laevis*, and *Sphaerium lacustre* are common in the Ipswichian, but scarce or absent in the Hoxnian, whilst the converse is true of *Gyraulus albus*. There is also some indication that the immigration of certain southern forms was relatively delayed, for example *Belgrandia* and *Corbicula*, both common in Ip II, are not known before Ho IIIa and IIIb respectively (Kerney 1971*a*). Furthermore, deposits at Swanscombe assigned to Ho III have yielded a distinctive suite of fluvial species with central or south-east European affinities which seems to have no analogy in the Ipswichian (†*Theodoxus serratiliniformis*, †*Viviparus diluvianus*, *Valvata naticina, Unio crassus*).

Cromerian faunas are known in any detail at three sites only (Sparks *in* Duigan 1963; Briggs *et al.* 1975; Kerney 1959*a*). But again, differences emerge from both the Ipswichian and the Hoxnian, e.g., in the absence of *Bithynia tentaculata* and *Corbicula*, and the presence of †*Nematurella runtoniana* and †*Valvata goldfussiana*. There is furthermore a hint that like the Hoxnian (but unlike the Ipswichian), maximum species diversity was not attained until Cr III (e.g., in the appearance of *Belgrandia, Valvata naticina*, and *Unio crassus*).

At a simple level of presence or absence Table 3.1 shows the known occurrence of certain freshwater species in the interglacials in southern England.

Turning to the terrestrial faunas, it is unfortunate that almost all our knowledge comes from water-laid sediments in which land snails (especially woodland snails) are under-represented. This is regrettable in view of the greater sensitivity of the latter to climatic and vegetational changes. Only for the Flandrian is a reasonably clear picture now emerging from a study of suitable subaerial sediments linked to carbon-14 dating. Fl I assemblages reflect the growth of open forest, with the appearance and expansion of *Carychium, Aegopinella, Vitrea, Lauria, Vertigo pusilla*, and *Discus ruderatus*. *Columella columella* persists. Important changes occur towards the end of Fl I, with the expansion of *Discus rotundatus* and *Oxychilus cellarius*, and the suppression of the remaining late-glacial open-ground elements. Fl II is characterized by the expansion of *Acicula fusca, Leiostyla anglica*, and *Spermodea lamellata*, imparting an oceanic facies. These changes form the basis of the series of molluscan assemblage zones (a) to (d) proposed by Kerney and Turner (1977). In Fl III human clearances are reflected by the re-expansion of *Vallonia*, and —mainly from Roman times—the appearance of introduced Mediterranean snails (*Helix aspersa, Monacha cantiana, Cernuella virgata*, etc.).

Table 3.1.
Known occurrence of certain freshwater snails in southern England

	Pre-Cr	Cr	Ho	Ip	Fl
†*Viviparus glacialis*	x				
Lithoglyphus naticoides	x				
†*Valvata goldfussiana*	x	x			
†*Pisidium clessini*	x	x	x	x	
Corbicula fluminalis	x		x	x	
†*Nematurella runtoniana*		x			
Valvata naticina		x	x		
Unio crassus		x	x		
Belgrandia marginata		x	x	x	
Bithynia inflata		x	x	x	
Marstoniopsis scholtzi		x		x	
Sphaerium solidum		x			?
Sphaerium rivicola		x	x		x
†*Theodoxus serratiliniformis*		?	x		
†*Viviparus diluvianus*			x		
Potomida littoralis			x	x	
Margaritifera auricularia				x	
Viviparus contectus				x	x
Theodoxus fluviatilis					x

Knowledge of pre-Flandrian interglacial land faunas remains much sketchier, especially for the Pre-temperate Zones (I), so that the pattern of development of the rich assemblages known from Zones II and III cannot be traced. On the other hand, in the case of the Ipswichian, Sparks (1964*b*) has argued for a real paucity of the Zone I fauna, immigration perhaps being delayed as a result of the exceptional severity of the Wolstonian glaciation. As with freshwater Mollusca, many thermophilous terrestrial species survived in reduced numbers throughout Ip IV.

Extra-British species occurring in the Ipswichian are *Vallonia enniensis* (south European), *Discus ruderatus* (boreo-continental), *Clausilia pumila*, *Bradybaena fruticum* (both central and east European), *Helicopsis striata* (central European), and †*Candidula crayfordensis*. Deposits of Ho II and III at Clacton and Swanscombe have yielded *Acicula polita*, *Macrogastra ventricosa* (both central European), *Vallonia enniensis*, *Discus ruderatus*, *Clausilia pumila*, and †*Candidula crayfordensis*; and calcareous tufas probably of Ho III date at Hitchin and Icklingham have additionally yielded †*Acicula diluviana*, *Semilimax semilimax*, *Clausilia parvula* (both central European), *Ruthenica filograna* (east European), †*Laminifera* sp.

(now Pyrenean), and †*Retinella* (*Lyrodiscus*) sp. (now Canaries). Few extra-British taxa are known from the Cromerian, and no extinct species: *Vallonia enniensis, Discus ruderatus, Semilimax semilimax, Clausilia pumila,* and *Trochoidea geyeri* (central European) are recorded.

Some conclusions

We are still very far from any detailed reconstruction of the history of non-marine molluscan faunas in the British Isles. Many further sequences must be studied, linked wherever possible to an independent system of dating. Some of the evidence is perhaps irrecoverably destroyed; for example, calcareous spring deposits, which give us our best insight into forest faunas, are scarcely known before the Flandrian. Furthermore, most of the sites so far examined lie in the south-east corner of the country, so that we have little knowledge of geographical variation within Britain at any one period.

As with flowering plants, there is no doubt that the pattern did not repeat exactly in each glacial–interglacial cycle. Doubtless a large element of chance is involved. Some of the observed differences may, however, reflect real climatic differences. For example, southern freshwater species are rather more strongly represented in the Ipswichian than in the Flandrian, Hoxnian, or Cromerian, suggesting higher mean temperatures. Also, the relative scarcity of *Lauria, Discus rotundatus,* and *Oxychilus,* coupled with the presence of non-British species of eastern range (notably *Bradybaena fruticum,* uniquely known from this interglacial) may reflect a climate more continental than in the Flandrian. Nevertheless, the Hoxnian reveals an even larger number of central European taxa, not easily reconcilable with the rather clear palaeobotanical evidence for a high degree of oceanicity.

Many fossil communities indeed at first sight seem to be made up of climatically 'incompatible' elements, and have no exact present-day counterparts. This is particularly true of periods of rapidly changing climate, such as the Late Devensian and early Flandrian, when thermophilous species with good powers of natural dispersal took advantage of newly created habitats to coexist with surviving cold elements. A further difficulty is that modern distributions may be misleading guides to true climatic requirements. A specially interesting case is provided by the two common European species of *Discus,* the boreo–alpine–continental *D. ruderatus,* and the west Mediterranean–Atlantic *D. rotundatus.* Their present geographical (or altimetric) segregation is compatible with their segregation in time in the English Flandrian record, but not with their frequent

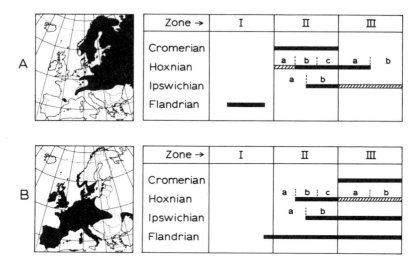

Fig. 3.1. Present distributions in Europe, and interglacial ranges in southern England, of (A) *Discus ruderatus* (Férussac) and (B) *D. rotundatus* (Müller). Solid bars = occurrences in deposits dated by pollen or radiocarbon; shaded bars = occurrences in deposits dated inferentially only.

association in the *middle* parts of interglacials (Fig. 3.1). Perhaps *D. ruderatus* has changed its thermal requirements, but it is equally likely that its present geographical range is not primarily controlled by climate. Unfortunately, direct studies of precise thermal tolerances analogous to the work of Iversen (1944) have scarcely been attempted for any European mollusc.

References

The ages of molluscs referred to in the papers below are indicated by appropriate Stage letter and Zone number. Stages: pC (pre-Cromerian); Cromerian: Anglian: Hoxnian: Wolstonian: Ipswichian: ED, MD, LD, Early, Middle, and Late Devensian: Flandrian.

Briggs, D.J., Coope, G.R. and Gilbertson, D.D. (1975). Late Pleistocene terrace deposits at Beckford, Worcestershire, England. *Geol. J.* 10, 1–16, (MD).

Briggs, D.J. and Gilbertson, D.D. (1973). The age of the Hanborough terrace of the River Evenlode, Oxfordshire. *Proc. Geol. Ass.* 84, 155–73, (W).

Briggs, D.J., Gilbertson, D.D., Goudie, A.S., Osborne, P.J., Osmaston, H.A., Pettit, M.E., Shotton, F.W. and Stuart, A.J. (1975). New interglacial site at Sugworth. *Nature, Lond.* 257, 477–79, (Cr III).

Burchell, J.P.T. and Piggott, S. (1939). Decorated prehistoric pottery from the bed of the Ebbsfleet, Northfleet, Kent. *Antiq. J.*, 19, 405–20.

Coope, G.R., Shotton, F.W. and Strachan, I. (1961). A Late Pleistocene fauna and flora from Upton Warren, Worcestershire. *Phil. Trans. R. Soc.* B 244, 379–421, (MD).

Cooper, J. (1972). Last Interglacial (Ipswichian) non-marine Mollusca from Aveley, Essex. *Essex Nat.* 33, 9–14, (Ip IIb–III).

Duigan, S.L. (1963). Pollen analyses of the Cromer Forest Bed Series in East Anglia. *Phil. Trans. R. Soc.* B 246, 149–202, (Cr II).

Evans, J.G. (1967). Late-glacial and post-glacial subaerial deposits at Pitstone, Buckinghamshire. *Proc. Geol. Ass.* 77, 347–64, (LD).

Evans, J.G. (1972). *Land snails in archaeology.* Seminar Press, London and New York.

Iversen, J. (1944). *Viscum, Hedera* and *Ilex* as climatic indicators. *Geol. Fören. Stockh. Förh.* 66, 463–83.

Kerney, M.P. (1959a). Pleistocene non-marine mollusca of the English interglacial deposits. Unpublished Ph.D. thesis, University of London.

Kerney, M.P. (1959b). An interglacial tufa near Hitchin, Hertfordshire. *Proc. Geol. Ass.* 70, 322–37, (Ho ? III).

Kerney, M.P. (1963). Late-glacial deposits on the chalk of south-east England. *Phil. Trans. R. Soc.* B 246, 203–54, (LD).

Kerney, M.P. (1968). Britain's fauna of land mollusca and its relation to the Post-glacial thermal optimum. *Symp. zool. Soc. Lond.* 22, 273–91.

Kerney, M.P. (1971a). Interglacial deposits in Barnfield Pit, Swanscombe, and their molluscan fauna. *J. geol. Soc. Lond.* 127, 69–93, (Ho IIb–III. W).

Kerney, M.P. (1971b). A Middle Weichselian deposit at Halling, Kent. *Proc. Geol. Ass.* 82, 1–12, (MD).

Kerney, M.P. (1976a). Mollusca from an interglacial tufa in East Anglia, with the description of a new species of *Lyrodiscus* Pilsbry (Gastropoda: Zonitidae). *J. Conch. Lond.* 29, 47–50, (Ho ? III).

Kerney, M.P. (1976b). A list of the fresh- and brackish-water mollusca of the British Isles. *J. Conch. Lond.* 29, 26–28.

Kerney, M.P., Brown, E.H. and Chandler, T.J. (1964). The Late-glacial and Post-glacial history of the chalk escarpment near Brook, Kent. *Phil. Trans. R. Soc.* B 248, 135–204, (LD. Fl I).

Kerney, M.P. and Turner, C. (1977). The biostratigraphy of Late Devensian and Flandrian deposits at Folkestone, Kent. *Proc. Geol. Ass.* 88, (Fl I–II).

Lambert, C.A., Pearson, R.G. and Sparks, B.W. (1963). A flora and fauna from late Pleistocene deposits at Sidgwick Avenue, Cambridge. *Proc. Linn. Soc. Lond.* 174, 13–29, (ED).

Large, N.F. and Sparks, B.W. (1961). The non-marine Mollusca of the Caincross Terrace near Stroud, Gloucestershire. *Geol. Mag.* 98, 423–26, (MD).

Morgan, A. (1969). A Pleistocene fauna and flora from Great Billing, Northamptonshire, England. *Opuscula Entomologica* 34, 109–29, (MD).

Morton, J. (1706). A letter to Dr. Hans Sloane, containing a relation of river and other shells digg'd up, together with various vegetable bodies, in a bituminous marshy earth, near Mears-Ashby, in Northamptonshire. *Phil. Trans. R. Soc.* 25, 2210–14.

Norris, A., Bartley, D.D. and Gaunt, G.D. (1971). An account of the deposit of shell marl at Burton Salmon, West Yorkshire. *The Naturalist* 917, 57–63, (Fl I–II).

Ovey, C.D. (ed.) (1964). The Swanscombe skull. A survey of research on a Pleistocene site. *Royal Anthropological Institute of Great Britain and Ireland.* Occasional Paper no.20, (Ho II–III).

Shotton, F.W. (1968). The Pleistocene succession around Brandon, Warwickshire. *Phil. Trans. R. Soc.* B 254, 387–400, (MD).

Sparks, B.W. (1957a). The non-marine mollusca of the interglacial deposits at Bobbitshole, Ipswich. *Phil. Trans. R. Soc.* B 241, 33–44, (**Ip I–II**).

Sparks, B.W. (1957b). The taele gravel near Thriplow, Cambridgeshire. *Geol. Mag.* 94, 194–200, (**W**).

Sparks, B.W. (1961). The ecological interpretation of Quaternary non-marine mollusca. *Proc. Linn. Soc. Lond.* 172, 71–80.

Sparks, B.W. (1962). Post-glacial mollusca from Hawes Water, Lancashire, illustrating some difficulties of interpretation. *J. Conch. Lond.* 25, 78–82, (**Fl I–II**).

Sparks, B.W. (1964a). A note on the Pleistocene deposit at Grantchester, Cambridgeshire. *Geol. Mag.* 101, 334–9, (**Ip ? III**).

Sparks, B.W. (1964b). The distribution of non-marine mollusca in the Last Interglacial in south-east England. *Proc. malac. Soc. Lond.* 36, 7–25.

Sparks, B.W. (1964c). Non-marine mollusca and Quaternary ecology. *J. Anim. Ecol.* 33 (Supplement), 87–98.

Sparks, B.W. and Lambert, C.A. (1961). The Post-glacial deposits at Apethorpe, Northamptonshire. *Proc. malac. Soc. Lond.* 34, 302–15, (**Fl I**).

Sparks, B.W. and West, R.G. (1959). The palaeoecology of the interglacial deposits at Histon Road, Cambridge. *Eiszeitalter Gegenw.* 10, 123–43, (**Ip III–IV**).

Sparks, B.W. and West, R.G. (1964). The interglacial deposits at Stutton, Suffolk. *Proc. Geol. Ass.* 74, 419–32, (**Ip III–IV**).

Sparks, B.W. and West, R.G. (1968). Interglacial deposits at Wortwell, Norfolk. *Geol. Mag.* 105, 471–81, (**Ip Ia, IIb**).

Sparks, B.W. and West, R.G. (1970). Late Pleistocene deposits at Wretton, Norfolk. I. Ipswichian Interglacial deposits. *Phil. Trans. R. Soc.* B 258, 1–30, (**Ip IIb–III**).

Sparks, B.W., West, R.G., Williams, R.B.G. and Ransom, M. (1969). Hoxnian Interglacial deposits near Hatfield, Herts. *Proc. Geol. Ass.* 80, 243–67, (**A, Ho I–II**).

Stelfox, A.W., Kuiper, J.G.J., McMillan, N.F., and Mitchell, G.F. (1972). The Late-glacial and Post-glacial Mollusca of the White Bog, Co. Down. *Proc. R. Ir. Acad.* B 72, 185–207, (**LD**).

Turner, C. and West, R.G. (1968). The subdivision and zonation of interglacial periods. *Eiszeitalter Gegenw.* 19, 93–101.

Waldén, H.W. (1976). A nomenclatural list of the land mollusca of the British Isles. *J. Conch. Lond.* 29, 21–25.

Wenz, W. and Zilch, A. (1938–60). Gastropoda (Prosobranchia and Euthyneura). *Handb. Paläozool.* 6, 1–2.

West, R.G. (1956). The Quaternary deposits at Hoxne, Suffolk. *Phil. Trans. R. Soc.* B 239, 265–356, (**A**).

West, R.G., Dickson, C.A., Catt, J.A., Weir, A.H. and Sparks, B.W. (1974). Late Pleistocene deposits at Wretton, Norfolk. II. Devensian deposits. *Phil. Trans. R. Soc.* B 267, 337–420, (**ED**).

West, R.G., Lambert, C.A. and Sparks, B.W. (1964). Interglacial deposits at Ilford, Essex. *Phil. Trans. R. Soc.* B 247, 185–212, (**Ip IIb**).

West, R.G. and Sparks, B.W. (1960). Coastal interglacial deposits of the English Channel. *Phil. Trans. R. Soc.* B 243, 95–133, (**Ip I–II**).

4

Marine Mollusca in the East Anglian Preglacial Pleistocene

P. E. P. NORTON

Abstract

Marine molluscan assemblages provide evidence of transport and derivation of the shells, water depth, and climate. Local marine facies, which may be correlated on a local scale, are recognizable, and summarized. Investigational methods are explained. Specific morphological studies (as of *Littorina* and *Spisula* in the Norwich district) and distributional studies (as of *Macoma balthica*) provide further insights, and are discussed. The chapter does not review stratigraphy of the deposits (see Chapter 18).

Investigation

Basis

The benthonic shell-bearing marine molluscs considered here are endofaunal or epifaunal loricates, gastropods, scaphopods, and bivalves. They are rather sedentary in life and associate with other animals and plants as described in the literature of marine animal communities reviewed by Thorson (1957). At death, their fossils may stay *in situ*, forming an autochthonous death assemblage, or they may be exhumed from sediment or detached from plants etc., and moved away to form transported (allochthonous) death assemblages. Both types contain many immature individuals, which does not prove that their species was unsuited to the habitat. Communities are not maintained by an indiscriminate larval rain, but by a somewhat 'targeted' supply of larvae. Each type of seabed has available a selection of larvae according to the breeding habits of the parents, the behaviour of the larvae, and oceanographic conditions (Thorson 1946). Mollusca are fossilized as occasional shells or as 'shell deposits' (allochthonous) or 'shell beds' (autochthonous death assemblages). *Broken shell deposits* contain shell fragments which have more of a sedimentary than paleoecological value, though they may have been formed by burrowing organisms, for example the shelly layers associated with *Arenicola* (Schäfer 1972). *Shell pavements* are deposits in which shells, exhumed from their original habitat, have been deposited with bivalves lying flat, open or closed, joined or separate (depending on the rate of accumulation of the cover sediments); the pavement may be simple or multilayered. *Shell beds* contain·shells which have remained substantially undisturbed since death and may be in the life position.

Quantitative examination of samples in the laboratory is expected to reveal in the death assemblage a spectrum of forms, local or from farther away, reflecting the general paleoecology of the area, the particular habitats found within it, the sedimentation processes, and so on, as discussed below. Morphological peculiarities of the shells may also be revealed, leading to paleoecological conclusions.

Method

Field samples of at least 2-4 kg are used (minimum 1 kg). An unaltered subsample is preserved. About 100 g are taken out for sediment analysis. Particulate samples are reduced to the desired weight for analysis (usually 1 kg) using a sample splitter ('Tyler' or similar). Clays are changed to a particulate condition by hydrogen peroxide treatment (West 1968). For ease of examination samples are sieved to 2000, 1000, 500, and 250 μm. Some shells pass the smallest sieve but no species is known to do it always, not even *Odostomia, Cae-cum*, or Skeneopsidae which do pass the 500 μm mesh. Contents of large gastropods are extracted. Even small fragments of shells can sometimes be identified, e.g. *Mytilus, Modiolus*, and *Pinna* which have characteristic crystal arrangements. Damage to shells may allow predation to be inferred, such as that caused by different drilling habits of *Natica* and *Nassarius*, and spire chopping by crabs (Schäfer 1972).

Numerical statements of frequency are used because they allow a wider range of interpretational techniques than verbal approxima-tions and their usefulness is less likely to go down into the grave with the author. In counting, one individual is defined as: a gastropod or its apex complete, two bivalve shells or their hinges complete (except *Anomia*, which is counted singly, as the right valve is not usually preserved); eight plates of a loricate; or so much of any of these that the remainder could not be identified by itself. Either all speci-mens in the sample are counted, or the first 600 in each fraction, after which the unexamined and examined material are weighed and the count multiplied up accordingly.

Interpretation

Basis

After the molluscs have been counted and the results tabulated, various statistical procedures are possible. The fossil assemblages are divided into groups, based on knowledge of their modern ecology and held to reflect the particular paleo-habitats found within the area. Where a succession of assemblages is available, changes in rela-tive frequency within and between the groups may be observed.

Marine thermal groups, supposed to reflect broad latitudinal distribution of the Mollusca which is controlled by temperature, have been used to deduce climatic changes (e.g. by Harmer 1902). Fossil assemblages are also divided into various 'ecological groups' of infauna, epifauna, predators, etc. which are also related to depth (e.g. Sorgenfrei 1958; Norton 1967 *et seq.*).

Derivation and transport

Death and deposition of molluscs are considered by Schäfer (1972). Insufficient work is available on the transporting of shells by water, e.g. results of flume tank experiments. Lever *et al.* (1964) carried out some experiments on an actual beach. Norton (unpublished) investi-gated the Eddystone shell gravel, an assemblage of live and dead shells forming in known conditions. It was found that shells showed little tendency to long-distance transport at 40 m depth. It appears to be impossible, at present, to measure the shells in a deposit, examine its composition, and produce a numerical reconstruction of the vectors concerned in its formation. The simpler criteria of autochthonousness already mentioned may be used if the deposit can be examined, but not for disturbed core samples. In attempting to deduce amounts of derivation and transport of an assemblage from laboratory counts only, some general principles are invoked. High percentages of badly worn shells (in practical terms, the element of the assemblage only identifiable to family or genus, etc) indicate rolling, transport, and degradation of the fossils, and thus allochthonous deposition. Low percentages of gastropod shells indicate unstable bottom conditions, due to turbulence or fast currents. Where mechanical analyses of the matrix can be compared with the records of sediment types inhabited by the contained species, further deduction may be possible. If such an analysis is to be valid, shells should not be removed from the sediment before particle size analysis, since they would (in the case of deposition of an assemblage *in situ*) be encountered by any settling larvae during the habitat selection stage. Where known granulometric tolerance of species and mechanical analysis of their matrix agree, there is a *possibility* of autochthonous deposition of the fossils.

Thus in the Ludham borehole (Norton 1967) the percentage of *Abra alba* increases from Zone LM 2 to LM 3. In Zone LM 2 silt clay percentages in the sediment rise from 5% to 25% and to 60% in Zone LM 3. Fine-sand percentages rise from 10% to 20% in Zone LM 2, and to 35% in Zone LM 3. *Abra alba* is considered by Eisma (1966) to live in seabeds with more than 50% of fine sand + silt + clay, including more than 5% of silt + clay. Ford (1923) gives the *Abra alba* community as inhabiting sediments with more than 50% of material <500 μm diameter—which obtains throughout most of Zones LM 2 and LM 3—including more than 20% of silt + clay (<60 μm diameter) —which obtains from -104 ft (31·7 m) upwards in Zone LM 2. Thus from that depth to the top of Zone LM 3 it is possible that *Abra alba* was deposited *in situ*, but this can only be confirmed by examination of undisturbed sediments (which are not available).

Water depth

The interpretation of marine facies (summarized below) in the East Anglian early Pleistocene is based primarily on estimations of water depth from Mollusca. There are many potential errors in any assessments of this, due to derivation and transport of the shells. Assemblages may be divided on grounds of known depth ranges of species, comparing frequencies of assumedly littoral, infralittoral, and sublittoral forms. Thus an assemblage containing, say, 80% of 'inter-tidal' forms, would be interpreted as having lived and been deposited in intertidal conditions, comparatively few 'sublittoral' shells being swept into the deposit. A high percentage of worn and indeterminable shells would also be expected unless indicators of sheltered tidal-flat conditions (e.g. *Hydrobia ulvae*) are frequent. Care is needed not to dogmatize the depth indication given by species, although it is natural for an ecologist familiar with the species in present-day waters to do so. It might be that the sea in which a fossil assemblage formed did not experience a tidal rhythm. Species can adapt to unusual circumstances, for example littoral species may 'submerge' in extremely low salinities. Thorson (1946) noted that *Mytilus edulis* and *Cardium edule* may be found living at 20 m deep or more in the inner Baltic where they occupy pockets of more saline water. Again, there are numerous *Littorina littorea* in the Netherlands 'Icenian' which suggest an intertidal rocky environment nearby, though Spaink (personal communication) considers that the early Pleistocene shoreline must have been tens of kilometres away to the south-east. The *Littorina* probably lived in clumps on open ground in 'wadden-area' situations. A last factor which may vitiate interpretations of water depth is the 'arctic emergence' shown by species, that is they live in shallower water as the northern limit of their range is approached. This is an imponderable question where no independent estimate of water temperature is available.

Interpretations of water depth from Mollusca have been the most important criteria in interpretation of marine facies in the pre-glacial Pleistocene of East Anglia. The types shown in Tables 4. 1 and 4. 2 have been recognized. For a fuller stratigraphic consideration, see Chapter 18.

Temperature, paleogeography

Data on thermal tolerances of molluscs that may be used for inter-preting paleotemperatures are scanty: see the review by Sorgenfrei (1958). Comparison of boreal, arctic, lusitanic, and such zoogeographical elements in fossil assemblages is indefinite. For example, in the Ludham borehole, Norton (1967, table 2) lists a 'sublittoral fine grade substratum infauna group'—*Abra alba, Macoma calcarea, Spisula subtruncata, Actaeon tornatilis, Cyprina islandica, Diplodonta astartea, Lucinoma borealis*, and *Serripes groenlandicus*—based on the faunal list for the whole 40 m of shelly sediments in the borehole. *M. calcarea* and *S. groenlandicus* are 'arctic', the rest 'boreal'. One expects that in cold periods these arctic species would increase their frequency in the assemblage while the others decrease. In fact their

Table 4.1.

Facies of Antian and Baventian Sites interpreted from Mollusca

Sizewell	Easton Bavents	Aldeby	Ludham	Sidestrand
Baventian, L 4b. Inner-sublittoral, muddy seabed with stony or shelly gravel. Worm tubes in sediments above this level (West and Norton 1974).	Baventian, L 4a, b. Inner-sublittoral perhaps deeper than Antian. Shell beds 1 (L 4a), 2 (L 4a).		Baventian, L 4b. Inner-sublittoral perhaps deeper water than Antian. Zone LM 6.	Baventian, not yet comparable with Ludham sequence. Open coast inter-tidal with impoverished assemblage and *Macoma balthica* SS/E, SS/K, S51 (Norton 1967).
	Antian, L 3. Inner-sublittoral, with impoverished assemblages. Shell beds 3, 4. Antian pollen between these two levels. ? early Antian, L 3. (acc. to zone boundaries of West (1961). Open coast facies. Shell bed 5 (Norton and Beck 1972).	?Antian Inner-sublittoral with diversity of habitats. (Baventian pollen found in clays above) (Norton and Beck 1972).	Antian L 3. Inner-sublittoral with estuarine littoral (Norton 1967).	

Table 4.2.
Pastonian facies and Bramerton facies interpreted from Mollusca

D. High-boreal or subarctic silty deposit facies with *Yoldia myalis, Mya truncata* (*in situ*). Chillesford Churchyard Pit 0·75 m above the level of the Chillesford Pollen assemblage; Aldeburgh Brickyard at same level as Chillesford Pollen assemblage (West and Norton 1974).

e) Open-coast, high-boreal or subarctic facies with littoral and sublittoral shells in the mollusc assemblage; Bramerton Common 5·80 m (Norton in press).

C. Open-coast facies succeeding 'B' (below), with *Littorina littorea, Cerastoderma edule, Hydrobia ulvae,* some *Spisula subtruncata,* and *Corbula gibba.* Wangford B1 and C, Sizewell 'B'9, possibly 'B'7. The Chillesford Pollen Spectrum occurred about 0·5 m below 'B'9 (West and Norton 1974).

c) Boreal 'fluviomarine' facies reflecting brackish sheltered tidal-flat conditions with *Hydrobia ulvae.* Blake's Pit W 1·40–1·55 m and Bramerton Common 9·80 m (Norton in press).

B. Sheltered estuarine facies, reflecting brackish conditions, with *Hydrobia ulvae.* Wangford B6 and B4, Sizewell 'B' nuclear power station roadway trench samples 1 and 4 (with Chillesford Pollen Assemblage), and Thorpe Aldringham (with Chillesford Pollen Assemblage) (West and Norton 1974).

b) Boreal littoral or inner-sublittoral facies with few brackish-water shells deposited, resembling (c) and (a) and could be regarded as intermediate. Blake's Pit E 2·71–2·81 m (Norton 1976).

a) Inner-sublittoral muddy-bottom facies with littoral, brackish, tidal-flat conditions represented; Bramerton Common 12·20 m (Norton in press).

['A' was listed West and Norton 1974 in sequence with 'B', 'C', and 'D', and described as 'sublittoral (or infralittoral) facies, shells abraded and not deposited *in situ.* Sizewell 'A' (pumphouse pit of the nuclear power station: pollen provisionally correlated with Ludham L4c, Baventian, occurred at this level), Chillesford '*Scrobicularia* Crag', possibly Aldeburgh '*Scrobicularia* Crag'. The Sizewell occurrence of this facies has already been mentioned.]

Note: The marginal letters are simply those used in the original publications. Neither they nor the juxtaposition of facies in this table are held to imply correlation between the Bramerton and Suffolk sequences.
The Chillesford Pollen assemblage has been provisionally correlated as Pastonian (West and Norton 1974).

highest frequencies (in Zone LM 1a, top and Zone LM 2, top, i.e. late in a temperate to cool and a cool to temperate succession respectively) are achieved without corresponding reduction in frequency of the other species. A much more detailed analysis was made by Valentine and Meade (1961) working on the Pleistocene of California. Norton (1967) attempted to find changes in Crag paleotemperatures by typing the molluscan assemblages as 'late-glacial', 'semi-interglacial', and 'full-interglacial' on the basis of lists of molluscs selected from the immigration sequence to the Osolfjord and Bohuslän clays and raised beaches (late- and post-Weichselian). A similar method was

used with apparent success by Funnell (1961) working on the Crag
Foraminifera. The Mollusca, however, gave no indication of cyclical
temperature change in the Ludham sequence. 'Boreal' molluscs (with
some 'arctic' and 'lusitanic' elements), were present during the temperate periods (Ludhamian, Antian) but there were no molluscs in
the sediments of the cool periods (Thurnian, Baventian). The Norwich Crag of Bramerton, and the Suffolk Pastonian deposit have a
similar 'boreal–mixed' fauna but the silts at Aldeburgh and Chillesford have a more arctic fauna (West and Norton 1974). Norton
(1970) suggested why the mollusc evidence differs from the pollen
and foraminiferal evidence. His assertion that mollusc associations
have a broader latitudinal range than vegetational or foraminiferal
associations of the same rank required checking. He later (1975)
proposed the existence of an early Pleistocene zoogeographic province with boundaries to the north of Iceland and to the south of
the Pentland Firth, in which molluscs were, by modern standards,
mainly 'boreal' with 'arctic' and 'lusitanic' admixtures. The northern
and southern forms could have been there because of peculiarities
of the thermal regime, or because of previously wider thermal tolerances of the species. The fauna is preserved in the East Anglia Crag
and in the Tjörnes deposits of Iceland, the 'boreal–mixed' characteristics being found in both; the similar anomalies at the two sites were
taken to indicate contemporary deposition within a wide-ranging
faunal province. This explanation is preferred to the alternative; that
the Icelandic deposit formed first, then the East Anglian one as the
(narrower) faunal province moved south. The North Sea was probably landlocked to the south in post-Ludhamian times (Funnell
1961), forming an inlet of the North Atlantic as the Baltic does now.
In temperate periods, molluscs from the main Atlantic moved into
the North Sea basin between Scotland and Norway, but show no
cyclical thermal variation (which the pollen and Foraminifera do)
due to the far-flung boundaries of their parent faunal province. In
cool times, there was development of silts with relatively cold-water
Mollusca at Chillesford and Aldeburgh (Pastonian), but with no
Mollusca, possibly due to eustatic lowering of sea-level, at Baventian
and Thurnian localities.

Oxygen-isotope paleotemperatures have not been found for the
Crags since the problems of correcting the results are at the moment
too great.

Summarizing the attempts made to interpret derivation and transport, water depth, temperature and paleogeography, it is apparent
that the sort of tentative conclusions reached are simplistic and will
no doubt be refined as comparisons and checking become possible.

According to Occam's Law of Economy, one must attempt to reach simple interpretations that are obviously suggested by the data.

Correlation

Microfossils such as pollen and Foraminifera, which give a regional interpretation and can be used to establish a chronology, have proved more useful than Mollusca for correlation purposes. Similar assemblages of Mollusca, indicating similar *facies* and occurring about 300 m apart laterally, have been correlated when comparing Bramerton Common Pit with Blake's Pit (Norton in press), but no other such cases have been put forward. Harmer and others were apparently satisfied that the Weybourne Crag at its various exposures could be correlated, but West and Wilson (1966) found it to be Baventian, Pastonian, and even Beestonian in age at different sites, and Norton (1970) considered it to be reworked at all save the Baventian sites. Thus similarity of facies is not held to indicate contemporaneity except in very local and well-understood situations. However, once the distribution of marine facies in the Crag basin is better understood on the basis of further malacological work (checked by pollen analysis), it may become possible to attempt more correlations. This will probably involve the recognition and comparison of Assemblage Zones based on the presently recognized series, and on the facies discussed above. Mollusca have also been used by other authors to establish Concurrent Range Zones and Acme Range Zones, which are abstractions; these are used in the zonation of the Netherlands Sequence (Spaink 1975) and were tried by Harmer (e.g. 1920). Additional collecting can alter the validity of such Zones; the amount of Netherlands material used to establish the Zones is, however, very large and detailed.

Evidence from shell morphology and distribution.

Although this chapter is concerned with analysis of mollusc death assemblages, valuable insights have been gained from specific morphological studies. Thus Cambridge (1975) has discussed some implications of the distorted *Littorina littorea* and *Thais lapillus* of Blake's Pit, Bramerton. Norton (in press) has added some suggestions and remarks on the aberrant *Spisula* specimens of the Arminghall pit and the Caistor St Edmunds (working) pit. These were noted by Harmer and Wood also; Harmer based an interpretation of deteriorating conditions and ice-approach on the many distorted specimens which he elevated to varieties. Parasitism, or more likely, enfeebling low-salinity or ice conditions together with settlement of barnacles on the shells may provide an explanation for the monstrosities commonly found in the gastropods. Others are keeled, this is thought to reflect stormy conditions. The *Spisula* forms are unusual and local, usually being labelled in collections as *Mactra solida* or *Mactra ovalis*. On re-examination, most proved to be *Spisula subtruncata* ('thick' type according to van Urk (1959)) with some *S. solida*, some *S. elliptica*, and one *S.* aff. *constricta* from Arminghall. The shells are misshapen and very large and thin. The presence of these forms is interpreted as an indication that the species lived in a restricted marine basin isolating them from competition by normal forms. The shell deposits most probably formed by the exhumation and re-deposition of patches of the aberrant forms. *Spisula subtruncata* today, for example, forms dense but limited patches, several square kilometres in extent and lasting only for one generation, in the North Sea (Davis 1923). The shell deposits may contain, therefore, only a few (even a single) generation of aberrant shells.

The problem of *Macoma balthica* distribution in the Crags has been the subject of several papers (Spaink and Norton 1966; Norton 1970; Norton and Spaink 1974). This highly competitive and successful species is absent from the Crag deposits except the Cromer Forest Bed Series and the Bure Valley Beds, where it is very rare and only juveniles are found (Norton in press). Norton has invoked the explanation of a late-Baventian transgression on the Cromer Forest Bed area, followed by extinction of the species locally in the suc-ceeding early Pastonian regression, and by a marine transgression later in the Pastonian when the Cromer Forest Bed received re-worked *Macoma balthica* shells (but no living populations of it) and the 'Norwich Crag' to the south received none of its shells at all. This idea remains without proof.

References

Beck, R.B., Funnell, B.M. and Lord, A. (1972). Correlation of Lower Pleistocene at depth in Suffolk. *Geol. Mag.* 109, 137–39.

Cambridge, P.G. (1975). Field meeting to Bramerton, near Norwich, 14–15 September 1974. *Bull. Geol. Assn. Norfolk.* 27, 33–46.

Davis, A.M. (1923). Quantitative studies on the fauna of the sea bottom. No. 1 Preliminary investigations of the Dogger Bank. *Min. of Agr. Fish. Invest. Ser.* 6(2).

Eisma, D. (1966). The distribution of benthic marine molluscs off the main Dutch coast. *Neth. J. Sea Res.* 3(1), 107–63.

Ford, E. (1923). Animal communities of the level sea-bottom in the waters adjacent to Plymouth. *J. mar. biol. Ass. U.K.* 13, 164.

Funnell, B.M. (1961). The Paleogene and Early Pleistocene of Norfolk. *Trans Norf. Norw. Natur. Soc.* 19(6), 340–65.

Funnell, B.M. and West, R.G. (1962). The Early Pleistocene of Easton Bavents, Suffolk. *Q. Jl. geol. Soc. Lond.* 117, 125–41.

Harmer, F.W. (1902). A sketch of the later Tertiary history of East Anglia. *Proc. Geol. Ass.* 7, 416–79.

Harmer, F.W. (1914–1925). *The Pliocene Mollusca of Great Britain.* (Includes Harmer 1920). Paleontographical Society, London.

Lever, J.J., van den Bosch, M., Cook, H., van Dijk, T., Thiadens, A.J.H. and Thijssen, R. (1964). Quantitative beach research: III. An experiment with artificial valves of *Donax vittatus. Neth. J. Sea Res.* 2(3), 455–92.

Norton, P.E.P. (1967). Marine molluscan assemblages in the early Pleistocene of Sidestrand, Bramerton and the Royal Society borehole at Ludham, Norrolk. *Phil. Trans. R. Soc.* B 253(784), 161–200.

Norton, P.E.P. (1970). The Crag Mollusca: a conspectus. *Bull. Soc. belge géol., Paléont. Hydrol.* 79(2), 157–66.

Norton, P.E.P. (1975). Paleoecology of the Mollusca of the Tjörnes sequence, Iceland. *Boreas* 4, 97–110.

Norton, P.E.P. (in press). Discussion of the depositional sequence of the Crag of the Norwich District as indicated by the Mollusca. *Mem. geol. Surv. U.K.* (New Series), Sheet 161.

Norton, P.E.P. and Beck, R.B., (1972). Lower Pleistocene molluscan assemblages and pollen from the Crag of Aldeby (Norfolk) and Easton Bavents (Suffolk). *Bull. geol. Soc. Norfolk* 22, 11–31.

Norton, P.E.P. and Spaink, G. (1974). The earliest occurrence of *Macoma balthica* (L.) as a fossil in the North Sea Deposits. *Malacologia* 14, 33–7.

Schäfer, W. (1972). *Ecology and paleoecology of marine environments.* Oliver & Boyd, Edinburgh.

Sorgenfrei, T. (1958). Molluscan assemblages from the Middle Marine Miocene of South Jutland and their environments. *Danm. geol. Unders.* II Raekke, 79.

Spaink, G. (1975). Zonering van het Mariene onder-Pleistoceen en Plioceen op grond van Mollusken-Fauna's. In *Toelichting bij geologische Overzichtskaarten van Nederland.* (ed. Zagwijn W.H. and C.J. van Staalduinen) Rijks Geol. Dienst, Haarlem.

Spaink, G. and Norton, P.E.P. (1967). The stratigraphical range of *Macoma balthica* (L.) [Bivalvia, Tellinacea] in the Pleistocene of the Netherlands and Eastern England. *Meded. Geol. Sticht. Nederlands* N.S. 18.

Thorson, G. (1946). Reproduction and larval development of Danish marine bottom invertebrates, with special reference to the planktonic larvae in the South (Øresund). *Meddr. Kommn Danm. Fisk. – og Havunders.* Ser. 4 Plankton, 4(1).

Thorson, G. (1957). Marine bottom communities. *Treatise on marine ecology and Paleoecology* (ed. J. Hedgpeth). *Mem. geol. Soc. Am.* 67.

Urk, R.M. van (1959). De Spisula's van het Nederlandse strand. *Basteria* 23, 1–29.

Valentine, J.W. and Meade, R.F. (1961). Californian Pleistocene paleotemperatures. *Univ. Calif. Publs. geol. Sci.* 40, 1–46.

West, R.G. (1968). *Pleistocene geology and biology.* Longmans, London.

West, R.G. (in press). The Cromer Forest Bed Series.

West, R.G. and Norton, P.E.P. (1974). The Icenian Crag of south-east Suffolk. *Phil. Trans. R. Soc.* (B) 269 (895), 1–28.

West, R.G. and Wilson, D.G. (1966). Cromer Forest Bed Series. *Nature, Lond.* 209, 497–98.

Wood, S.V. (1848–1882). *The Crag Mollusca.* Paleontographical Society, London.

5

Quaternary Coleoptera as aids in the interpretation of environmental history

G. R. COOPE

Abstract

Studies of Quarternary fossil Coleoptera have shown that there is little evidence of any evolution, either morphological or physiological, during this period. Assemblages of fossil beetles provide much environmental information on glacial and interglacial episodes but, as yet, they cannot be used for stratigraphical correlation. The rapidity and intensity of their response to changes in the thermal environment is well demonstrated by the faunas from the latest glacial period. During some warmer interludes, of interstadial status, assemblages of temperate beetles inhabited the British Isles whilst the flora of the times lacked trees. Such biotal imbalance is interpreted primarily as a function of time—the more mobile components of the biota being able to take advantage of any climatic amelioration well before the more pedestrian members of the flora could respond.

Introduction

The systematic investigation of fossil insect assemblages in Quaternary environmental studies is a relatively recent development compared with the attention that has been devoted to vertebrates, molluscs, or to micro- and macro-botanical material. Sporadic efforts were made in the first half of this century to identify and interpret Quaternary insect fossils but these to a large extent foundered upon two serious obstacles. First there was a widely held view that insect species had evolved in relatively recent times and would certainly be expected to differ considerably from their modern representatives even within the period of the latest interglacial–glacial cycle. The result of this attitude of mind was that fossils were credited with a wealth of new names, some of which conveyed little or no information of any pre-sumed relationship with any modern species (Scudder 1895). The frequent use of 'pre' before the trivial name implied that the fossils represented ancestral forms of known modern species. This nomen-clatural procedure had unforeseen repercussions, for example Lom-nicki (1894) described a fossil water beetle from the Starunia 'woolly rhinoceros site' as *Helophorus prenanus*, an unfortunate name since Angus (1973) has shown that the species is clearly the same as *H. jacutus*, (described by Poppius (1907)) which today lives in eastern Siberia. The fossil is certainly *not* ancestral to the european *H. nanus*. Unfortunately Lomnicki's name was given before that of Poppius and thus has precedence, perpetuating the name and the mis-understanding upon which it was based. What is doubly unfortunate is that this species of *Helophorus* was a frequent member of the Quaternary fauna of the British Isles.

The second obstacle in the path of the earlier Quaternary entomo-logists was the inadequacy of geological information. The immense complexity of Quaternary terrestrial stratigraphy meant that it was often difficult to allocate the fossil-bearing sites to any unequivocal stratigraphical position, and the frequency and rapidity of environ-mental changes that we now know to have occurred could hardly then have been guessed at. Thus Henriksen in his monumental work on the Quaternary insect faunas of Denmark and Scania (Henriksen 1933) recognized that the species he was dealing with were identical in almost all cases with their modern representatives. However, the stratigraphical framework in which he places his finds is too broad to be of much value today.

Even Henricksen found difficulty in taking into account *post mortem* changes and erected a new species *Notiophilus coriaceous* on the basis of differences that Lindroth (1948) has demonstrated to be the results of the process of fossilization. Lindroth points out the degree to which an expert can be misled into faulty identification by the unfamiliarity of species represented by disarticulated skeletons or by the rather bizarre features that are the results of *post mortem* changes. These changes take the form of distortion in shape, the development of spurious punctures or more rarely of tubercles, a peculiar rugosity on the surface of some individuals, and finally differences in colour of the fossils compared with their modern representatives. Fortunately many of these alterations appear only in the dried specimens and can thus be recognized and allowed for if the fossils are first investigated under alcohol.

It has become abundantly clear that the fossil insects, particularly the Coleoptera, show no evidence of any morphological change during the latter half of the Quaternary. So far, the only evidence of any evolutionary change has been found by Matthews (1974) amongst early Pleistocene fossils from Alaska. The differences are indeed small and of the same order of magnitude as the variations that we see today between geographical races of a single species. However, Matthews (1976) has shown that in faunas of Pliocene age from the Beaufort formation in arctic Canada, there are species which show differences from their modern equivalents on the same scale as those that distinguish closely related species at the present day. These preliminary results suggest that, for the Coleoptera of temperate and northern latitudes at least, speciation took place for the most part during the Upper Tertiary. It is possible, even likely, that this high degree of evolutionary stability may not be applicable to island situations remote from the possibility of frequent replenishment from the mainland, and to isolated caves between which species transference is all but impossible. Under such circumstances atypically rapid evolution might occur, at rates far higher than under more usual conditions.

The demonstrable morphological stability of species of Coleoptera during the Quaternary is an essential prerequisite if we are to use them as environmental indicators. If the fossils had shown any progressive difference from their modern equivalents, this would almost certainly have been associated with physiological change which altered their environmental tolerances. Although physiological stability cannot be demonstrated experimentally, it seems from the way that species kept the same company in the past as they do today, that they must have shared common environmental requirements then as now. The degree to which well-marked communities of species can be recognized certainly as far back as the Cromerian Interglacial, is a measure of this physiological stability at the species level.

Fossil insects occur in a wide variety of deposits. It has been known for a long time that they often occur in peats as conspicuously coloured semi-articulated skeletons. However, the felted nature of the peat makes the extraction of the fossils difficult other than by visual scanning of the bedding planes, which results in an assemblage biased in favour of the most conspicuous species (Coope 1961). Much better results are obtained from organic silts. In these the insect fossils are often beautifully preserved and easily recovered by wet sieving, resulting in a more representative sample of the fossils in the deposit. Lake muds, especially if they are collected near the margin, also produce abundant Coleoptera, but it is necessary to obtain large-diameter (minimum 10 cm) cores. As a general rule, insect fossils can be found in any sediment that contains macroscopic plant remains.

The context of the fossiliferous sediment will indicate to a large extent the size and nature of the area supplying the fossil sample. Thus a deposit that accumulated on the flood plain of a river in a landscape devoid of trees is likely to contain species washed or blown into it from a wide area, say several square kilometres. In a deposit representing the litter from a forest floor the area sampled may be very much smaller. In some archaeological contexts the fossil assemblage may represent the population that lived and died within the confines of a single room. Because many beetles resort to flight as a dispersal mechanism, they may also stray well away from suitable habitats before becoming incorporated in a deposit. As a general rule, in the environmental interpretation of Quaternary insect assemblages it must be borne in mind that in most cases a wide spectrum of available habitats will be represented. This will limit the precision with which one can reconstruct the local conditions but, apart from some archaeological contexts where the environment of a few square metres may be of interest, the Coleoptera can provide a remarkably intricate picture of the habitats available in the neighbourhood.

As a group the Coleoptera inhabit an extremely wide range of habitats, though most species are restricted to relatively narrow environmental choices. They have colonized almost every terrestrial and fresh-water situation and some are even characteristic of the marine intertidal zone. They exist at all trophic levels in the community; some are specialized phytophages restricted to a small range of food plants while others find many species of host plants acceptable; some are fungus eaters and others feed upon decaying vegetation itself. There are numerous carnivores, either specialized or generalized predators that may eat one another, molluscs, or even small vertebrates such as fish and small frogs. In many cases their food chains extend down through the soil arthropod fauna to the Colembola and mites that feed on algae, lichens, and fungi. They may thus be independent of the macrophytes that make up the bulk of the flora recorded in palynological diagrams. (This explains why many beetles, in colonizing newly available habitats, do not have to wait upon the growth of the flowering plants, but can move in as soon as lichens and algae can become established). Some beetle species occupy specialized habitats such as dung or carrion, whilst others live in mammal burrows, the nests of bees or wasps, or abandoned bird nests. Some are aquatic, for either all or part of their life history, and may be characteristic of running or standing water, richly or poorly supplied with nutrients. Terrestrial beetles may provide information on the degree of vegetation cover or the nature of the soil, whether it be clay, sand or gravel, damp or dry, saline or not, poor or rich in humus. Indeed, the precision of beetle species in their choice of living conditions is often very exact and the summary given above is a very inadequate account of their range of habitat preferences.

Perhaps the most significant contribution that Quaternary Coleoptera can make to the understanding of past environments is, however, in the field of climatic interpretation. All entomologists would agree that the most fundamental environmental factor governing the geographical distribution of a species is the climate and in particular the thermal environment, measured either as intensity in terms of temperature or as available heat expressed in day–degrees. Notwithstanding the intricacies of arguments about the precise factors which affect single species, it is evident that for many their distributions today correspond with well-known climatic zones. Assuming that these distributions are in balance with the climatic conditions, and not merely reflections of the stage reached by a particular species in extending its range into a new habitat, it is reasonable to interpret those whose distributions are narrowly restricted to a climatic zone as stenothermic species, whilst those of broader range may be considered as eurytherms and thus of less importance in palaeoclimatic interpretation. Of course the range of a species of beetle may be determined by factors other than climate, such as the availability of a food plant or some specialized habitat, and so it is of the utmost importance to understand as much as possible about the modern ecology of the species before embarking upon an assessment of its climatic significance.

The recent expansion of the study of Quaternary entomology has revealed two important and initially unexpected aspects of the response of Coleoptera to climatic changes. The first of these is the enormous scale of the changes in geographical distribution of species in response to changes of climate. Thus, within the period of the last glacial–interglacial cycle alone, some species have shifted the limits of their geographical distributions by as much as 7000 km (Ullrich and Coope 1974). Another indication of the readiness of beetle species to move in response to climatic changes may be recognized by the fact that, even as late as 'Younger Dryas' time, up to 40% of the species in some of our assemblages have subsequently become extinct in the British Isles. The second aspect of the response of the Coleoptera to climatic change is the extreme rapidity with which the specific composition of a fauna may change. Temperate faunas replace arctic assemblages and vice versa within a few centimetres of pond bottom sediment. This combination of sensitivity and rapidity of response to climatic changes, coupled with their demonstrable evolutionary stability, makes the Coleoptera one of the most climatically significant components of the whole terrestrial biota.

Interglacial insect faunas

Assemblages of fossil insects have been obtained from deposits of the last three interglacials so far recognized in Britain; the Cromerian, Hoxnian, and Ipswichian. These faunas often include species whose present-day ranges come no nearer to these islands than southern or central Europe, which makes the identification of these fossils a difficult and time-consuming procedure. This is because the number of species that have to be taken into consideration in our efforts to identify the fossils increases almost exponentially towards the south of Europe compared with the relatively impoverished north (Britain has only about 3700 species of beetle). Interglacial insect fossils are also difficult to study because the deposits themselves are often well humified and compressed, making the recovery of the fossils in acceptably large fragments frequently impossible. In spite of these problems, interglacial deposits that contain silt or fine sand yield preserved fossils that on occasions prove to be as unexpected members of the fauna of the British Isles as the hippopotamus with which they were associated (for example the tropical genus *Drepanocerus* in the deposits under Trafalgar Square, London). In many ways, therefore, the study of interglacial insect faunas is in a very embryonic condition compared with the investigation of fully glacial or interstadial assemblages, and the discussion that follows will have to be drawn from the few published accounts and also much as yet unpublished data.

One of the most outstanding characteristics of interglacial fossil assemblages is the way in which different components of the biota are in harmony in providing a consistent picture of the local environment (contrast with the situation in some interstadials, p. 64). This environmental harmony is well illustrated by the floral and faunal sequences from the Nechells Interglacial (Hoxnian) described respectively by Kelly (1964) and by Shotton and Osborne (1965). Here the changes in the insect fauna could be tied into the pollen diagram and the varying representation of species of Mollusca, Polyzoa, ostracods, and fish. An example of this correlative behaviour may illustrate this point.

Rhynchaenus quercus is a weevil that feeds almost exclusively upon oak and its occurrence in the Nechells deposits matches, both in stratigraphic range and in individual frequency, the spectrum for *Quercus* pollen.

Although the Nechells deposits are so far the only ones to have their sequential biotal changes compared, a considerable number of other interglacial deposits, each representing only a brief time span, have been investigated and the interpretations of the local environmental conditions are in complete agreement regardless of which part of the biota is employed. Thus in the Sugworth (Cromerian) deposits, fossil plants, molluscs, vertebrates, and insects present a picture of a river meandering through a largely deciduous woodland with open glades and areas of marsh along its banks (Briggs *et al.* 1975). The Austerfield (Ipswichian) site yielded a flora and insect fauna (Gaunt *et al.* 1972) that suggests that the deposit accumulated in a dystrophic pool fed by small streams and surrounded by marshy ground with abundant sedge. Around this pool were forests dominated by *Carpinus*, an interpretation based on botanical data but supported by the relative abundance of the southern European *Scolytus carpini* which lives principally on *Carpinus*. The classical site at Ipswich has produced abundant botanical evidence (West 1957) and an assemblage of Coleoptera (Coope 1975), both of which suggest almost identical local environmental conditions, namely a rather eutrophic lake partially overgrown with reeds and surrounded by lightly wooded country. The abundance of dung beetles is amply supported by recent excavations in the neighbourhood by Wymer (personal communication) which have produced an extensive suite of large mammal bones from contemporary deposits. The unpublished information on the insects from the Ipswichian interglacial deposits of Trafalgar Square along with that of the plant fossils, vertebrates, and molluscs, presents a picture that precisely resembles that from the type site at Ipswich. It has been beautifully portrayed by the late Neave Parker in the Illustrated London News (Franks *et al.* 1958).

Taken as a whole, these consistent pictures of the local environments of all available interglacial deposits strongly support the
assumption of insect physiological stability that was made earlier
(p. 57). They also suggest that most of the components of a fossil
assemblage were derived from the local communities of plants and
animals that lived in the neighbourhood rather than being an assort-
ment of species brought together either actively or passively from a
variety of distant habitats.

Some idea of the climatic conditions that prevailed at the time
when a deposit was being laid down may be obtained by a consideration of the present-day ranges of the species found as fossils within
it. All the three mentioned interglacial periods provide deposits that
yield fossils of insect species whose ranges lie entirely to the south
of these islands and, as far as I am aware, no species whose ranges
lie exclusively to the north. Before considering specific examples,
however, it is important to recognize that the activities of man in
modern times may have modified considerably the effective range
of a species chiefly by restriction of suitable habitats. In no case is
this modification likely to be more pronounced than in clearance of
the old forest and removal of mature and dead timber in the interests
of forest hygiene. Fossil beetles from postglacial sites in Britain include species which have been forced to abandon this country in
more modern times, more probably as a result of forest clearance
than because of any climatic change (Buckland and Kenward 1973).

A substantial proportion of the beetle fauna from the Sugworth
(Cromerian) interglacial deposits are species that live exclusively to
the south of Britain; *Oxytelus opacus, Pelochares versicolor*, and
Valgus hemipterus, and the list of these southern species is still being
added to as work progresses on the fauna. There can be no doubt
that this interglacial had considerably warmer summer temperatures
than those in Britain today (Briggs *et al.* 1975). The Nechells (Hoxnian) Interglacial deposits yielded a beetle fauna that 'suggests that
the climatic optimum [of the interglacial] represents conditions
not very different from those of [Britain] today' (Shotton and
Osborne 1965). Only three species occurred that are no longer living
in Britain, and of these one is an uncertain species of *Ataenia* and
the other two are bark beetles whose distribution today may be
modified by human activities. The beetle fauna currently being investigated by the author from the type site at Hoxne itself supports
the general climatic conclusions drawn from the Nechells fossils. In
the assemblage of Coleoptera from the type section of the Ipswichian
Interglacial, 29% of the species are no longer living in Britain and
have modern ranges chiefly to the south. These include *Oodes graci-*

lis, Cybister lateralimarginalis, Airaphilus elongatus, Caccobius schreberi, Onthophagus opacicollis, and *Valgus hemipterus*. As a whole the fauna suggests warm summers with average July temperature of 20°C, which is about 3°C above that for southern England at the present time. There is no reason to believe that the winter temperatures were any colder than they are in the same area today. A large assemblage of Coleoptera from the Trafalgar Square interglacial deposits also contains a high proportion of southern and central European species and suggests a similarly warm climate.

In summary, the fossil insect assemblages from the three interglacials suggests that, at their 'climatic optima' the Cromerian and Ipswichian periods were considerably warmer than the present time and almost certainly warmer than the present interglacial reached at its thermal maximum. The Hoxnian interglacial seems to have been no warmer than the present day.

Correlation of interglacial deposits by means of their insect faunas is at the moment premature because so few have been adequately investigated. However, this is not the only problem. The specific composition of insect faunas of very different ages can often be astonishingly similar in spite of the large number of species concerned and the differences that might be expected from random errors in sampling. Two examples may be noted. The lists of species from the Nechells (Hoxnian) Interglacial deposits (Shotton and Osborne 1965) and from the Austerfield (Ipswichian) Interglacial deposits (Gaunt *et al.* 1972) are very similar indeed and might tempt the unwary to equate the two deposits. Palynologically, however, the two sites must be allocated to separate interglacials. The close similarity of the insect faunas is most probably a reflection of similarity of the local environment rather than identity of the age of the deposits. Another example of similar faunas of very different ages is seen in the lists from the two incompletely investigated sites at Sugworth (Cromerian) and Trafalgar Square (Ipswichian). Not only are many of the exotic species common to the two assemblages but a number of distinctive, but as yet unidentified, specimens also occur at both sites. However, the stratigraphical setting and the vertebrate and molluscan faunas from the two sites forbid any correlation of these deposits and once again, the close similarity of the beetle faunas must reflect similarity of local habitats.

Since interglacial insect faunas reflect so closely the immediate environment in which the deposits accumulated, it is to be expected that the variations in the numerical representation of the species in time, after the manner of pollen diagrams, will be indicative of

changes in the local environments and thus be of value in stratigraphical correlation only in a very limited area.

Insect faunas of the last (Devensian, Weichselian) glaciation

Fossil insect assemblages from the various phases of the most recent
glacial period are more completely understood than are those of the
interglacials and at the time of writing (early 1976) insect-bearing
deposits from 43 sites have been investigated. Several recent papers
review the current position, particularly with respect to the climatic
implications of coleopteran assemblages (Coope 1975).

In contrast to the situation in interglacials, the biota of glacial
periods does not always show that its various components were in
balance with the environment. This is particularly so after periods
of rapid but intense climatic amelioration when in the scramble to
occupy newly available habitats, the more mobile and less environ-
mentally fastidious elements of the flora and fauna would be the
pioneer settlers, way ahead of their pedestrian but perhaps ultimately
more efficient competitors. Under these circumstances the composition of a fossil assemblage is not to be equated with modern floras
and faunas that have had ample time since the retreat of the ice
sheets to establish balanced communities. It is this role of time—
time for the ecosystem to reach equilibrium—that I believe has been
underestimated as an essential factor in the interpretation of periods
of rapid climatic changes.

For the most part the flora and fauna of the last glaciation are in
complete accord in their climatic implications. Thus the insect fauna
from Chelford Interstadial at its type site (Coope 1959) and the flora
described by West (Simpson and West 1958) provide independent
evidence for an almost identical reconstruction of the local environment and climate. Similarly the evidence of both the flora and fauna
throughout the colder phases of the glaciation also present a unified
picture of widespread tundras in Britain with a somewhat continental
climate (Coope 1968; Morgan 1973; Bell 1969).

However, there are two outstanding periods when floral and faunal
evidence appear to offer contrasting environmental interpretations.
These apparent conflicts are entirely confined to inferences about
the climate of the times, since both flora and fauna present identical
views of the local environment. During the thermal maximum of the
Upton Warren Interstadial, a thermophilous assemblage of Coleoptera was associated with a completely treeless landscape (Coope and
Angus 1975). The Coleoptera suggest a typically west European temperate climate and they provide no evidence for believing in a climate
regime at this time equivalent to that of the steppe. It seems most

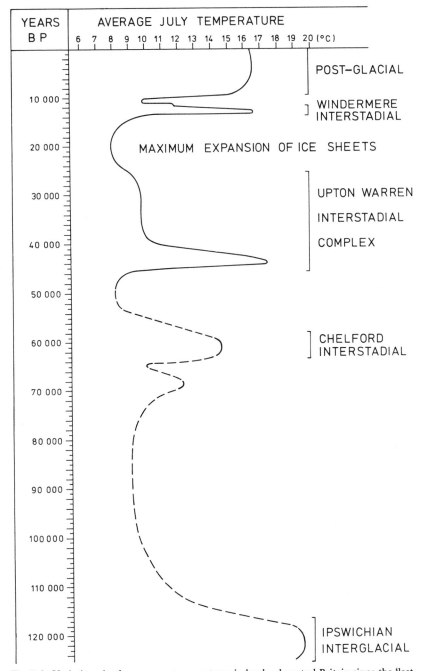

Fig. 5.1. Variations in the summer temperatures in lowland central Britain since the 'last interglacial', based on evidence of the fossil Coleoptera.

reasonable to interpret the absence of trees in terms of history rather than the climate of the times. After a rapid amelioration of climate, the trees had simply not had enough time to reach Britain from their refuges in southern Europe (see Fig. 5. 1).

A very similar situation prevails in the Windermere Interstadial (Coope and Pennington 1977) ('Late glacial interstadial') when the insect fauna shows abundant evidence of a thermal maximum as warm as the present day before the development of birch wood-land in Britain, for long taken as the index of climatic amelioration. By the time that the birch woodland had become well established in these islands (equivalent in time to the Allerød) the climate had deteriorated to such an extent that the assemblages of Coleoptera and the floral evidence once again are in complete environmental accord, indicating climates considerably cooler than those in low-land Britain today (Ashworth 1972).

With the close of the glacial period there was a short episode of biotal inbalance as the sudden and intense recovery in the thermal environment once again started a rush to colonize the new lands of the north. This time the race started so much closer to our southern shores that the disparity in arrival times was much less than at the start of the Windermere Interstadial. Thus by 9500 years ago, a thermophilous assemblage of Coleoptera, that should by rights have inhabited a mixed oak forest, lived in the English Midlands in a woodland of birches (Osborn 1974). At a similar time in south west Scotland the birch forest was there associated with species of beetles, indicating climatic conditions as warm as or even warmer than those of the present day (Bishop and Coope 1977).

References

Angus, R.B. (1973). Pleistocene *Helophorus* (Coleoptera Hydrophilidae) from Borislav and Starunia in the Western Ukraine with a reinterpretation of M. Łømnicki's species, description of a new Siberian species, and comparison with British Weichselian faunas. *Phil. Trans. R. Soc.* B 265, 299–326.

Ashworth, A.C. (1972). A Late-glacial fauna from Red Moss, Lancashire, England. *Entomol. scand* 3, 211–24.

Bell, F.G. (1969). The occurrence of Southern, Steppe and Halophyte elements in Weichselian (Last-Glacial) floras from Southern Britain. *New Phytol.* 68, 913–22.

Bishop, W.W. and Coope, G.R. (1977). The environmental history of Late glacial and early Postglacial times in south west Scotland. In *Studies in the Scottish Late glacial environment*. (ed. J.M. Gray and J.J. Lowe), Pergamon, London.

Briggs, D.J., Coope G.R. and Gilbertson, D.D. (1975). Late Pleistocene terrace deposits at Beckford, Worcestershire, England. *Geol. J.* 10, 1–16.

Buckland, P.C. and Kenward, H.K. (1973). Thorne moor: a paleo-ecological study of a bronze-age site. *Nature, Lond.* 241, 405-06.

Coope, G.R. (1959). A Late Pleistocene insect fauna from Chelford, Cheshire. *Proc. R. Soc. Lond.* B 151, 70-86.

Coope, G.R. (1961). On the study of glacial and interglacial insect faunas. *Proc. Linn. Soc. Lond.* 172, 62-5.

Coope, G.R. (1968). An insect fauna from Mid-Weichselian deposits at Brandon, Warwickshire. *Phil. Trans. R. Soc.* B 254, 425-56.

Coope, G.R. (1975). Climatic fluctuations in north-west Europe since the Last Interglacial, indicated by fossil assemblages of Coleoptera. In *Ice Ages ancient and modern.* (ed. A.E. Wright and F. Moseley), *Geol. J. Spec. Issue* 6, Liverpool.

Coope, G.R. and Angus, R.B. (1975). An ecological study of a temperate interlude in the middle of the Last Glaciation, based on fossil coleoptera from Isleworth, Middlesex. *J. Anim. Ecol.* 44, 365-91.

Coope, G.R. and Brophy, J.A. (1972). Late glacial environmental changes indicated by a coleopteran succession from North Wales. *Boreas* 1, 97-142.

Coope, G.R. and Pennington, W. (1977). The Windermere Interstadial of the Late Devensian. *Phil. Trans. R. Soc.* B 208.

Franks, J.W., Sutcliffe A.J., Kerney, M.P. and Coope, G.R. (1958). Haunt of elephant and hippopotamus: the Trafalgar Square of 100 000 years ago—new discoveries. *The Illustrated London News*, 1011-13.

Gaunt, G.D., Coope, G.R., Osborne P.J. and Franks, J.W. (1972). The interglacial deposit near Austerfield, Southern Yorkshire. *Inst. Geol. Sci. report* 72/4, 1-13.

Henriksen, K. (1933). Undersogelser over Danmark—Skanes Kuartaere Insectfauna. *Vidensk. Meddr. dansk. naturh. Foren.* 96, 77-355.

Kelly, M.R. (1964). The Middle Pleistocene of North Birmingham. *Phil. Trans. R. Soc.* B 247, 533-92.

Lindroth, C.H. (1948). Interglacial insect remains from Sweden. *Sver. geol. Unders.* 42, 1-29.

Lomnicki, A.M. (1894). Pleistocenskii owady z Boryslawia. (Fauna Pleistocenica insectorum Borislaviensium). *Wydaw. Muz. Dzieduszyck.* 4, 3-116.

Matthews, J.V. (1974). Quaternary environments at Cape Deceit (Seward Peninsula, Alaska): evolution of a tundra ecosystem. *Bull. geol. Soc. Am.* 85, 1353-84.

Matthews, J.V. (1976). Evolution of the subgenus Cyphelophorus (Genus *Helophorus*, Hydrophilidae, Coleoptera). Description of two new fossil species and discussion of *Helophorus tuberculatus* Gyll. *Can. J. Zool.* 54, 652-76.

Morgan, A. (1973). Late Pleistocene environmental changes indicated by fossil insect faunas of the English Midlands. *Boreas* 2, 173-212.

Osborne, P.J. (1972). Insect faunas of Late Devensian and Flandrian age from Church Stretton, Shropshire. *Phil. Trans. R. Soc.* 263, 327-67.

Osborne, P.J. (1974). An insect assemblage of Early Flandrian age from Lea Marston, Warwickshire, and its bearing on the contemporary climate and ecology. *Quaternary Res.* 4, 471-86.

Poppius, B. (1907). Beitrag zur Kenntnis der Coleopteren—fauna des Lena-Thales in Ost Siberien 3. *Ofvers. K. VetenskAkad. Förh.* 49, 1-17.

Scudder, S. (1895). Coleoptera hitherto found fossil in Canada. *Bull. geol. Surv. Can.* 2, 27-56.

Shotton, F.W. and Osborne, P.J. (1965). The fauna of the Hoxnian Interglacial deposits from Nechells, Birmingham. *Phil. Trans. R. Soc.* B 248, 353-78.

Simpson, I.M. and West, R.G. (1958). On the stratigraphy and palaeobotany of a late-Pleistocene organic deposit at Chelford, Cheshire. *New Phytol.* 57, 239–50.

Ullrich, W.G. and Coope, G.R. (1974). Occurrence of the east palaearctic beetle *Tachinus jacuticus* Poppius (Col. Staphylinidae) in deposits of the Last glacial period in England. *J. Ent.* (B) 42, 207–12.

West, R.G. (1957). Interglacial deposits at Bobbitshole, Ipswich. *Phil. Trans. R. Soc.* B 241, 1–31.

6

British Quaternary vertebrates

A. J. STUART

Abstract

It is now possible to assign many fossil vertebrate records to particular stages, substages, and zones, and where material has been found in association with pollen, the fauna and vegetation can be related. Significant advances have been made in unravelling the faunal history of the early Middle Pleistocene, e.g. the fauna of the type Cromerian (s.s.) of West Runton has been described, and a cave or fissure assemblage of probable Cromerian age has been discovered at Westbury. Much vertebrate material of Hoxnian age was found during recent archaeological excavations at Hoxne, Clacton, and Swanscombe. Small mammals have been described from probable Wolstonian, Ipswichian, and Devensian horizons at Tornewton Cave. Ipswichian vertebrates are sufficiently well known for changes in the fauna in relation to climatic and vegetational changes to be followed through the stage. Some Devensian and Flandrian material has been recently discovered. A promising beginning has also been made on detailed studies of particular taxa, notably squirrels, beavers, and lagomorphs.

Introduction

Fossil vertebrates, mostly mammals, from Quaternary deposits of the British Isles have attracted the attention of workers since early in the nineteenth century. The emergence in the last few years of a firm stratigraphical framework for the Quaternary deposits of the British Isles (see Mitchell *et al.* 1973) has given considerable impetus to modern studies of the fossil vertebrates. Attention has focused mainly on the faunas of the early Middle Pleistocene, Hoxnian, and Ipswichian stages, while a significant start has been made on the detailed history of particular taxa using biometrical techniques.

Many of the major fossil localities, including those mentioned in the chapter, are given in Fig. 6. 1.

Faunal history

Vertebrate fossils occur mainly in fluviatile and cave deposits, but they have also been found in marine and lacustrine sediments. Lower Pleistocene (marine 'Crags') and early Middle Pleistocene (Cromer Forest Bed Series) deposits are mostly confined to East Anglia. The Anglian fauna is almost unknown. No records of Pre-Devensian age are available for either Scotland or Ireland. Most cave faunas are of Upper Pleistocene age except for one dating from the early Middle Pleistocene and one from the Lower Pleistocene. An outline faunal history of the British Pleistocene is given by Stuart (1974).

Lower Pleistocene (Waltonian to Baventian)

Vertebrate fossils from the crags (Fig. 6.1) include fishes, marine mammals, and sparse terrestrial vertebrates. Many can only be assigned broad age limits, due both to poor stratigraphical control of many finds and the problems of relating much of the crag deposits to the type sequences. The vertebrates have received relatively little attention since Newton's (1891) study. Spencer (1964) has stressed the distinction between the fauna of the Nodule Bed (e.g. Woodbridge, Suffolk) and that of the overlying Red Crag.

Carreck (1966) described a vole, *Mimomys pliocaenicus*, from the Baventian of Easton Bavents, Suffolk. A few large mammals from here can be assigned to separate Antian and Baventian horizons (Stuart 1974).

Kurten's (1968) book includes identifications of the crag Carnivora, and *Hypolagus brachygnathus*, a lagomorph new to the British Pleistocene, has been recorded from the Norwich Crag of Bramerton, Norfolk (Mayhew 1975).

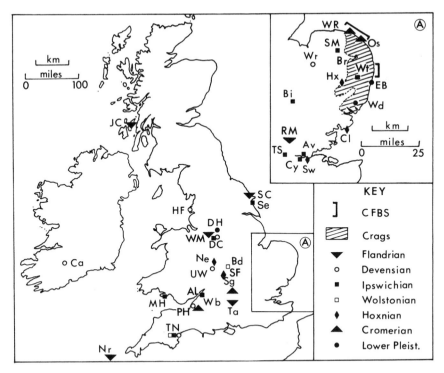

Fig. 6.1. Location Map. The extent of the crag deposits (mainly Lower Pleistocene) and the cliff sections in the Cromer Forest Bed Series (early Middle Pleistocene) are indicated. Key to site symbols:
Al, Alveston (fissure); Av, Aveley; Bd, Brandon, Warwickshire; Bi, Barrington; Br, Bramerton; Ca, Castlepook Cave; Cl, Clacton; Cy, Crayford area; DC, Derbyshire and Staffordshire caves; DH, Dove Holes (cave or fissure); EB, Easton Bavents; HF, High Furlong, Blackpool; Hx, Hoxne, JC, Jura caves; MH, Minchin Hole and Bacon's Hole (caves); Ne, Nechells, Birmingham; Nr, Nornour, Scilly Isles; Os, Ostend; near Bacton; PH, Picken's Hole (cave); RM, Rammey Marsh; SC, Star Carr; Se, Sewerby; SF, Stretton-on-Fosse; Sg, Sugworth, near Oxford; SM, Swanton Morley and Beetley; Sw, Swanscombe and Ingress Vale; Ta, Thatcham; TN, Tornewton Cave; TS, Trafalgar Square, London; UW, Upton Warren; Wb, Westbury (cave or fissure); Wd, Woodbridge; WM, White Moor, Bosley; WR, West Runton; Wr, Wretton, Wt, Wortwell.

The fauna from the cave or fissure deposit at Dove Holes, Derbyshire (Dawkins, 1903) has been reinvestigated by Spencer and Melville (1974).

Early Middle Pleistocene (Pastonian to early Anglian)
The Cromer Forest Bed Series (CFBS), long famous for its vertebrate fauna, is exposed along the coast of Norfolk and Suffolk (Fig. 6.1). The deposits have often in the past been referred to loosely as

'Cromerian', but are now regarded as representing a number of stages of which the Cromerian (s.s.) is only one (West and Wilson 1966). Much of the abundant fossil material in museum collections lacks the information necessary to relate it to one or other of these stages.

Important older works on the CBFS faunas include Newton (1882, 1891), Hinton (1926) and Azzaroli (1953) and more recently McWilliams (1967) has described many finds.

The rich vertebrate fauna of the Upper Freshwater Bed (Zone Cr II) of the type Cromerian (s.s.) site at West Runton, Norfolk, has recently been re-examined (Stuart 1975) and related to vegetational conditions. The general picture is of a slow-flowing river with aquatic plants, and also fishes, amphibians including frogs and/or toads waterbirds, and mammals from waterside habitats such as *Desmana moschata* (Russian desman), *Castor fiber* (beaver), *Trogontherium cuvieri* (extinct beaver), and *Mimomys savini* (a vole). A regional temperate forest supported *Apodemus* (wood, mouse, *sylvaticus* group), *Clethrionomys glareolus* (bank vole), *Dicerorhinus etruscus* (extinct rhinoceros), *Sus scrofa* (wild boar), and *Capreolus capreolus* (roe deer). Local herb-dominated vegetation on the river flood-plain appears to have supported grassland voles, mostly *Microtus* cf. *arvalis* (common vole) and other animals which probably avoided dense forest, e.g. *Equus* sp., *Libralces* (=*Alces*) *latifrons* (an elk) and *Premegaceros* (=*Megaloceros*) *verticornis* (a giant deer). The assemblage appears to have accompanied largely terrestrial material washed into the channel. A sparse fauna from the over-lying marine gravel at West Runton ('Monkey Gravel' of Hinton (1908), of Zone Cr III age, includes *M. savini* together with *Macaca* sp. (a macaque), and *Scirus whitei* (a squirrel), taxa not recorded from Zone Cr II (Stuart 1975).

Sediments from Ostend, Norfolk, from which C. Green collected vertebrates in the early nineteenth century, have been referred by West to pollen Zone Cr IV (Stuart and West 1976). The fauna includes *Arvicola cantiana* as the only water vole present, indicating that it replaced *Mimomys savini* at about the Cr III–IV boundary. This information is valuable in assessing the age of other sites in Britain and the continent which have faunas of Cromerian type, but with *Arvicola* instead of *Mimomys*.

A fauna, probably Pastonian, from a series of marine sands at East Runton, Norfolk, includes *Mimomys newtoni* (a small species), a large species of *Mimomys*, *Euctenoceros tetraceros*, and *Dama nestii* (extinct deer), but lacks many taxa present in the Cromerian (Stuart 1974). Recent excavations for sea defences at West Runton exposed similar sediments of undoubted Pastonian age (R.G. West, personal communication) beneath the Upper Freshwater Bed. A few small-vertebrate remains including *M. newtoni* have already been discovered in this deposit.

In 1969 a cave or fissure deposit containing very large numbers of vertebrate fossils was discovered in a quarry in Carboniferous Limestone near Westbury-sub-Mendip, Somerset. M.J. Bishop is engaged on a stratigraphical and faunal study and has published his preliminary findings (Bishop 1974, 1975), The oldest horizon, Bed 1 (Siliceous Group), comprises sands and gravels with a sparse fauna, mostly a small species of *Bison* together with *Hyaena brevirostris* (extinct hyaena). Bishop considers that this fauna is no younger than late Lower Pleistocene. Beds 2 to 9 (Calcareous Group), consist of limestone breccias and conglomerates and have yielded much *Ursus deningeri* (extinct bear), plus *Felis gombaszoegensis* (extinct leopard), *Xenocyon lycanoides* (extinct dhole), both new records in the British Pleistocene, *D. etruscus*, and human artefacts. Bed 10 (Rodent Earth), contains mainly *Microtus* spp. plus *Pitymys gregaloides* (extinct pine vole), *Arvicola cantiana, Lemmus* and *Dicrostonyx* (lemmings), *Ochotona* (pika), and *Pliomys episcopalis* (extinct vole), another new record. The faunas of Beds 2-10 are broadly comparable with those of the CFBS as a whole and with those loosely described as 'Cromerian' from continental Europe. The placing of the Ostend fauna in Zone Cr IV (see above) suggests that much of the Westbury sequence may be of similar age.

Hoxnian

New material has been discovered at several Hoxnian sites, largely as a result of archaeological investigations.

The fluviatile deposits of the Clacton Channel, mainly Zone Ho IIb (Turner and Kerney, in Kerney 1971) have yielded a number of mammals including *Palaeoloxodon antiquus* (straight-tusked elephant), *Equus caballus, Dama dama* (fallow deer), and *Bos primigenius* (aurochs) (Sutcliffe 1964). Recent excavations at the Clacton Golf Course site have produced a modest fauna including *Trogontherium cuvieri* (Singer *et al.* 1973). However, these authors consider that this horizon pre-dates Zone Ho I.

Barnfield Pit, Swanscombe, Kent where the famous human skull was found, has yielded numerous mammal remains from a succession of horizons (Sutcliffe 1964). The fauna of the Lower Loam is very like that of the Clacton Channel, while those of the later horizons show a decrease in woodland elements and the Silt Bed has yielded remains of *Lemmus lemmus*. Recent excavations in the Lower Loam directed by Waechter have produced a number of vertebrates at present being studied by Sutcliffe and Currant. They include *Arvicola cantiana, Pitymys* sp. (Sutcliffe and Kowalski, 1976), cf. *Macaca* and *Talpa* (mole). *Oryctolagus cuniculus* (rabbit) has been recorded by Mayhew (1975). Of great interest is the horizon of footprints, mainly made by ungulates, which have been skilfully preserved for future study.

Excavations in nearby Dierden's pit, Ingress Vale, Kent, directed by Sieveking, have yielded a few vertebrates in addition to those previously recorded from here (see Sutcliffe 1964).

Excavations at the type site at Hoxne, Suffolk, directed by Wymer have resulted in the discovery of much vertebrate material, where little was known previously. The fauna, at present being studied by Singer and Wolff (Gladfelter and Singer 1975; Singer personal communication) appears to be very similar to those of other Hoxnian sites, with the important addition of *Macaca* sp. and *Desmana moschata*. As at Swanscombe, *L. lemmus* occurs in the upper part of the sequence which is perhaps of Ho IV age.

A fauna of probable Hoxnian age was described by Shotton (1973) from Stretton-on-Fosse, Warwickshire.

Several fishes are recorded from the Hoxnian lake deposits at Nechells, Birmingham (Shotton and Osborne 1965).

Wolstonian

Wolstonian vertebrates are rather rare and have been little studied. Faunas of probable Wolstonian age are recorded from the Glutton and Bear Strata of Tornewton Cave, Devon (Sutcliffe and Zeuner 1962). Rodents from these horizons (Kowalski 1967) include *Cricetus cricetus* (common hamster), *Phodopus songorus* (dwarf hamster), *Lagurus lagurus* (steppe lemming), *Dicrostonyx torquatus, Lemmus lemmus*, and various voles.

A number of freshwater fishes were recorded from a Wolstonian channel deposit at Brandon, Warwickshire (Osborne and Shotton 1968).

Ipswichian

At Aveley, Essex in 1964 two superimposed elephant skeletons were found in fluviatile organic sediments. The lower (*Palaeoloxodon antiquus*) lay in sediment of Zone Ip IIb age, whereas the upper (*Mammuthus primigenius*) lay in Zone Ip III (West 1969). They are now displayed as excavated in the British Museum (Natural History). Remains of other vertebrates were also recovered.

Mammal remains have recently been found at Swanton Morley and Beetley, Norfolk in fluviatile sediments of Zone Ip IIb and Ip III age (Phillips 1972). A partial skeleton of *P. antiquus* was found in Zone Ip IIb sediments at Wortwell, Norfolk (Sparks and West 1968).

A rich well-preserved assemblage was recovered from Barrington, Cambridgeshire, around the turn of the century. Pollen analyses from sediment adhering to bones from this site have demonstrated that the deposit and fauna are of Zone Ip IIb age (Gibbard and Stuart 1975). The very high levels of herb pollen (about 90%) suggest that the river floodplain was extensively deforested, although the regional vegetation was mixed oak forest. The fauna includes probable forest animals, e.g. *Palaeoloxodon antiquus, Dama dama*, together with *Crocuta crocuta, Panthera leo* (lion), *Dicerorhinus hemitoechus* (a rhinoceros), *Hippopotamus amphibius*, and *Megaloceros giganteus* (giant deer), which probably preferred areas of open herbaceous vegetation near the river. These areas may have been maintained and extended by the activities of herbivorous mammals.

The history of the vertebrate fauna in relation to vegetational changes through the stage is discussed by Stuart (1976). The tundra-like vegetation of the preceeding Wolstonian cold stage, with a 'steppe–tundra fauna' (see above), gave way to regional mixed oak forest and a temperature fauna, including *P. antiquus, D. dama*, and H. amphibius, by Zone Ip IIb in response to ameliorating climatic conditions with warm summers and mild winters. Locally deforested areas in many river valleys supported such animals as *D. hemitoechus* and *M. giganteus* (as at Barrington, see above). At the beginning of Zone Ip III the forest appears to have become generally more open and *M. primigenius* and Equus caballus reappeared. *H. amphibius* is last recorded from early in this zone. Towards the end of Zone Ip IV (Crayford) open boreal forest appears to have been established, accompanied by a fauna transitional to that of the Devensian with *Ovibos moschatus* (musk ox), *Coelodonta antiquitatis* (woolly rhinoceros), and lemmings. That the Ipswichian climate was generally warmer that now is supported by the occurrence of *Emys orbicularis* (pond tortoise), *Hippopotamus*, and *Crocidura* cf. *suaveolens* (lesser white toothed shrew—a new record from Aveley, Zone Ip IIb), all well north of their present ranges. On the other hand, the presence of animals of tundra—*Dicrostonyx torquatus, Lemmus lemmus, O. moschatus*; and steppe—*Spermophilus undulatus* (Siberian long tailed souslik) (see Mayhew 1975) at Crayford appears to reflect generally open vegetational conditions rather than a cold climate.

Boylan (1967) has revised the fauna of the Sewerby–Hessle deposits, Yorkshire. Taylor (1973) has recorded mammals of probable Ipswichian age from a fissure at Alveston, Avon, and Bramwell

(1964) recovered remains of both Ipswichian and Devensian vertebrates from Elderbush Cave, Staffordshire.

Kowalski (1967) listed rodent records from the Hyaena stratum of
Tornewton Cave. Rzebik (1968) described material of a large species
of *Crocidura* (not previously recorded from mainland Britain) from
the 'Vivian Vault' in this cave.

Stringer (1975) has published a preliminary account of the stratigraphy and palaeontology of Bacon Hole Cave. Much of the fauna
was identified by Currant, the rodents by Stuart. The rodents show
an interesting sequence with the boreal *Microtus oeconomus* (northern vole) at the base and top, with temperate taxa between. Excavations at nearby Minchin Hole by Sutcliffe and Bowen have also
resulted in the recovery of faunal remains. The small vertebrates
from the Ipswichian horizons at Wretton, Norfolk have been studied
by Carreck, and Ipswichian vertebrates have been found in recent
excavations directed by Wymer at Stoke Tunnel, Ipswich.

Devensian

Comparatively little work has been done recently on Devensian vertebrates. Material from a series of Early Devensian terrace deposits at
Wretton, Norfolk is being studied by K.A. Joysey. His preliminary
faunal list (personal communication) includes *Alopex lagopus* (arctic
fox), *Mammuthus primigenius, Equus caballus, Rangifer tarandus*
(reindeer), and *Bison* sp.

An important fauna of Middle Devensian age comes from Upton
Warren, Worcestershire (Coope, Shotton, and Strachan 1961). The
mammals comprised *M. primigenius, C. antiquitatis, Bison, R. tarandus, E. caballus, Dicrostonyx torquatus*, and *Microtus* sp. Many cave
deposits contain faunas of probable Early and Middle Devensian age,
e.g. Picken's Hole, (Mendip Hills) Somerset (Stuart 1974). The Elk
and Reindeer strata in Tornewton Cave, Devon produced abundant
large mammal remains (Sutcliffe and Zeuner 1962) and rodents including *Dicrostonyx torquatus, Microtus oeconomus*, and *M. gregalis*
(tundra vole) (Kowalski 1967). The fauna from Castlepook Cave,
County Cork, Ireland has recently been revised by Sutcliffe (in
preparation). The Irish Pleistocene fauna as a whole has been reviewed
by Savage (1966) and discussed by Mitchell (1969). The faunas are
markedly impoverished compared with those of England. The little-
known Scottish faunas have been reviewed by Delair (1969).

Late Devensian finds from England include a near-complete *Alces
alces* (elk) skeleton from Zone II lacustrine deposits at High Furlong,
Blackpool, Lancashire (Hallam *et al.* 1973). Records of birds, together with other vertebrates, have been listed from a number of cave

deposits of Late-Devensian age by D. Bramwell. The fauna from a single horizon in Ossom's Cave, Manifold Valley, Staffordshire, includes birds, e.g. *Bubo bubo* (eagle owl), *Lagopus mutus* (ptarmigan), *Lyrurus tetrix* (black grouse), *Pluvialis apricaria* (golden plover), *Corvus corax* (raven), and mammals, e.g. *R. tarandus, E. caballus, D. torquatus, L. lemmus, M. oeconomus*, (Bramwell personal communication). A radiocarbon date of 10 950 ± 70 years BP was obtained for this horizon. A rather similar fauna has been recorded from Thor's Cave in the same valley.

Flandrian

Flandrian vertebrates have been rather neglected in recent years, in spite of their obvious relevance to studies of the modern fauna.

From early Flandrian Zone Fl I, mammals and birds are recorded from archaeological sites at Star Carr, Yorkshire (Clark 1954) and Thatcham, Berkshire, (Wymer, 1962); a few small-vertebrate remains from Ramney Marsh, North London (Stuart 1974); and fishes and amphibians from lake marl at Bosley, Cheshire (Johnson, Franks, and Pollard 1970).

Microtus oeconomus, no longer found in Britain, and other small mammals have been identified by Pernetta and Handford (1970) from Bronze and Iron age deposits on Nornour, Isles of Scilly.

A number of Flandrian bird faunas have been identified by D. Bramwell (in preparation). The sites include Dowel and Fox Hole Cave, Earl Sterndale, and Demen's Dale Rock Shelter, Taddington, all in Derbyshire.

Studies of particular taxa

A detailed study of British Quaternary beavers, squirrels, and lagomorphs has been made by Mayhew (1975) applying knowledge of variation in recent populations to fossil material.

The absolute individual ages of a large sample of subfossil Flandrian beavers (*Castor fiber*) were determined by the examination of cement layers in the cheek teeth. This has formed the basis for investigation of ontogenetic changes and estimation of population characters in the sample.

Castor fiber shows little change through the Pleistocene except for fluctuations in size. Pastonian animals are considerably larger than those from the Cromerian and other stages, but there is no overall trend of size change with time. *Trogontherium*, on the other hand, shows a progressive size increase from the early Pleistocene to the Cromerian, then slight diminution in size, followed by its extinction in the Hoxnian.

Re-investigation of the fossil ground squirrels has shown that they can be referred to extant species: *Spermophilus undulatus* (early Anglian, late Ipswichian (Crayford)) and *S. major* (Devensian). Sufficient Crayford material was available for survivorship curves to be constructed based on tooth wear classes.

Lepus timidus from Upper Pleistocene cave sequences shows marked fluctuations in size, with the large forms dating from the Wolstonian and Devensian cold stages, consistent with a southward shifting of the modern size cline of about 20 degrees. The applications of such size changes to stratigraphical interpretation are emphasized.

The British Pleistocene rodents are reviewed by Sutcliffe and Kowalski (1976). The paper includes comprehensive lists of specimens and localities together with taxonomic revisions.

Biometrical studies of Flandrian and recent bones of water fowl are being made by E.M. Northcote and investigations of British Pleistocene *Ursus maritimus* (polar bear) and *Gulo gulo* (glutton) have been made by Kurten (1968, 1973).

Included in the account of the type Cromerian fauna (Stuart 1975) are simple biometrical data for *Talpa* sp. (mole) and voles. Pasquier (1972) had previously shown that only one species of *Mimomys* is present at this horizon.

Biometrical data for Ipswichian voles are presented as part of a study of the faunal history of this stage (Stuart 1976). Biometrical study of the fossil rhinoceros material indicates that *D. hemitoechus* was the only species of *Dicerorhinus* present during this interglacial. Histograms of lamellar indices for pollen-dated elephant molars allow partial distinction of *Palaeoloxodon antiquus* from *Mammuthus primigenius*.

Two very interesting studies which have been published demonstrating relatively short-term changes in voles, mainly during the Flandrian, point the way to future work. Both used material from archaeological horizons in caves. Montgomery (1975) studied *Arvicola terrestris* (water vole) material from Derbyshire caves ranging from the Late Devensian to about 2000 years BP together with recent material from Shropshire. He found an overall increase in size and decrease in the degree of pro-odonty with time. Corbet (1975) looked at *Microtus agrestis* material from a cave on the Isle of Jura, Scotland, covering the period from pre-2500 years BP to the present day and found progressive changes in dental character with time.

Acknowledgments

I am grateful to a number of colleagues who have freely supplied information on their current research, in particular M.J. Bishop, D. Bramwell, A. Currant, K.A. Joysey, D.F. Mayhew, R. Singer, C. Stringer, and A.J. Sutcliffe.

References

Azzaroli, A. (1953). The deer of the Weybourne Crag and Forest Bed of Norfolk. *Bull. Br. Mus. nat. Hist.* (A. Geology). 2, 3–96.

Bishop, M.J., (1974). A preliminary report on the Middle Pleistocene mammal-bearing deposits of Westbury-Sub-Mendip, Somerset. *Proc. speleol. Soc.* 13, 301–18.

Bishop, M.J. (1975). Earliest record of man's presence in Britain. *Nature, Lond.* 253, 95–7.

Boylan, P.J. (1967). The Pleistocene mammalia of the Sewerby-Hessle buried cliff East Yorkshire. *Proc. Yorks. geol. Soc.* 38, 175–200.

Branwell, D. (1964). The excavations at Elderbush Cave, Wetton, Staffs. *N. Staffs. J. Field Stud.* 4, 46–50.

Briggs, D.J., Gilbertson, D.D., Goudie, A.S., Osborne, P.J., Osmaston, H.A., Pettit, M.E., Shotton, F.W. and Stuart, A.J. (1975). New interglacial site at Sugworth. *Nature, Lond.* 257, 477–9.

Carreck, J.N. (1966). Microtine remains from the Norwich Crag (Lower Pleistocene) of Easton Bavents, Suffolk. *Proc. Geol. Ass.* 77, 491–6.

Clark, J.G.D.C. (1954). *Excavations at Star Carr.* University Press, Cambridge.

Coope, G.R., Shotton, F.W. and Strachan, I. (1961). A late Pleistocene fauna and flora from Upton Warren, Worcestershire. *Phil. Trans. R. Soc.* B 244, 379–421.

Corbet, G.B. (1975). Examples of short- and long-term changes of dental pattern in Scottish voles (Rodentia; Microtinae). *Mammal Rev.* 5, 17–21.

Dawkins, W.B. (1903). On the discovery of an ossiferous cavern of Pleistocene age at Dove Holes, Buxton (Derbyshire). *Quart. Jl. geol. Soc. Lond.* 59, 105–33.

Delair, J.B. (1969). North of the hippopotamus belt: a brief review of Scottish fossil mammals. *Bull. Mammal Soc. Br. Isl.* 31, 16–21.

Gibbard, P.L. and Stuart, A.J. (1975). Flora and vertebrate fauna of the Barrington Beds. *Geol. Mag.* 112, 493–501.

Gladfelter, B.G. and Singer, R. (1975). Implications of East Anglian glacial stratigraphy for the British Lower Palaeolithic. In *Quaternary Studies* (eds. R.P. Suggate, M.M. Cresswell). The Royal Society of New Zealand, pp. 139–45.

Hallam, J.S., Edwards, B.J.N., Barnes, B. and Stuart, A.J. (1973). A late glacial elk with associated barbed points from High Furlong, Lancashire. *Proc. prehist. Soc.* 39, 100–28.

Hinton, M.A.C. (1908). Note on the discovery of the bone of a monkey in the Norfolk Forest Bed. *Geol. Mag.* 5, 440–4.

Hinton, M.A.C. (1926). *Monograph of the voles and lemmings. (Microtinae), living and extinct.* British Museum (Natural History), London.

Johnson, R.H., Franks, J.W. and Pollard, J.E. (1970). Some Holocene faunal and floral remains in the Whitemoor meltwater channel at Bosley, East Cheshire. *N. Staffs. J. Field Stud.* 10, 65–74.

Kerney, M.P. (1971). Interglacial deposits in Barnfield Pit, Swanscombe and their molluscan fauna. *Jl. geol. Soc. Lond.* 127, 69–93.

Kowalski, K. (1967). *Lagurus lagurus* (Pallas 1773) and *Cricetus cricetus* (L. 1758) (Rodentia, Mammalia) in the Pleistocene of England. *Acta zool. cracov.* 12, 111–22.

Kurtén, B. (1964). The evolution of the polar bear *Ursus maritimus* Phipps. *Acta zool. fenn.* 108, 1–26.

Kurtén, B. (1968). *Pleistocene mammals of Europe*. Weidenfeld and Nicolson, London.

Kurtén, B. (1973). Fossil glutton (*Gulo gulo* (L.)) from Tornewton Cave, South Devon. *Commentat. biol.* 66, 1–8.

McWilliams, B. (1967). *Fossil vertebrates of the Cromer Forest Bed in Norwich Castle museum*. Modern Press, Norwich.

Mayhew, D.F. (1975). The Quaternary history of some British rodents and lagomorphs. Thesis, University of Cambridge.

Mitchell, G.F. (1969). Pleistocene mammals in Ireland. *Bull. Mammal Soc. Br. Isl.* 31, 21–5.

Mitchell, G.F., Penny, L.F., Shotton, F.W., and West, R.G. (1973). A correlation of Quaternary deposits in the British Isles. *Geol. Soc. Lond. Spec. Rep.* 4.

Montgomery, W.I. (1975). On the relationship between sub-fossil and recent British water voles. *Mammal. Rev.* 5, 23–9.

Newton, E.T. (1882). The vertebrata of the Forest Bed Series of Norfolk and Suffolk. *Mem. geol. Surv. U.K.*

Newton, E.T. (1891). The vertebrata of the Pliocene deposits of Britain. *Mem. geol. Surv. U.K.*

Osborne, P.J. and Shotton, F.W. (1968). The fauna of the channel deposit of early Saalian age at Brandon, Warwickshire. *Phil. Trans. R. Soc.* B 254, 417–24.

Pasquier, L. (1972). Etude d'une population de *Mimomys savini* Hinton 1910 (Arvicolina, Rodentia) provenant de l'Upper Freshwater Bed (Quaternarie ancien d'Angleterre). *Mammalia* 36, 214–25.

Pernetta, J.C. and Handford, P.T. (1970). Mammalian and avian remains from possible Bronze Age deposits on Nornour, Isles of Scilly. *J. Zool. Lond.* 162, 534–40.

Phillips, L. (1972). Pleistocene vegetational history and geology in Norfolk. Thesis, University of Cambridge.

Rzebik, B. (1968). *Crocidura* Wagler and other Insectivora (Mammalia) from the Quaternary deposits at Tornewton Cave in England. *Acta zool. cracov.* 13, 251–63.

Savage, R.J.G. (1966). Irish Pleistocene mammals. *Ir. Nat. J.* 15, 117–30.

Shotton, F.W. (1973). A mammalian fauna from the Stretton Sand at Stretton-on-Fosse, South Warwickshire, *Geol. Mag.* 109, 473–6.

Shotton, F.W. and Osborne, P.J. (1965). The fauna of the Hoxnian interglacial deposits of Nechells, Birmingham. *Phil. Trans. R. Soc.* B 248, 353–78.

Singer, R., Wymer, J.J., Gladfelter, B.G. and Wolff, R.G. (1973). Excavation of the Clactonian industry at the Golf Course, Clacton-on-Sea, Essex. *Proc. prehist. Soc.* 39, 100–28.

Sparks, B.W. and West, R.G. (1968). Interglacial deposits at Wortwell, Norfolk. *Geol. Mag.* 105, 471–81.

Spencer, H.E.P. (1964). The contemporary mammalian fossils of the Crags. *Trans. Suffolk Nat. Soc.* 12, 333–44.

Spencer, H.E.P. and Melville, R.V. (1974). The Pleistocene mammalian fauna of Dove Holes, Derbyshire. *Bull. geol. Surv. Gt. Br.* 48, 43–53.

Stringer, C. (1975). A preliminary report on new excavations at Bacon Hole Cave, Gower. *Gower* 26, 32–7.

Stuart, A.J. (1974). Pleistocene history of the British vertebrate fauna. *Biol. Rev.* 49, 225–66.

Stuart, A.J. (1975). The vertebrate fauna of the type Cromerian. *Boreas* 4, 63–76.

Stuart, A.J. (1976). The history of the mammal fauna during the Ipswichian/ Last Interglacial in England. *Phil. Trans. R. Soc.* B 276, 221–50.

Stuart, A.J. and West, R.G. (1976). Late Cromerian fauna and flora at Ostend, Norfolk. *Geol. Mag.* 1, 469–73.

Sutcliffe, A.J. (1964). In: The Swanscombe skull. A survey of research on a Pleistocene site. (ed. C.D. Ovey). *Occ. Pap. R. Anthrop. Inst.* 20.

Sutcliffe, A.J. and Kowalski, K. (1976). Pleistocene rodents of the British Isles. *Bull. Br. Mus. nat. Hist. (Geol.)* 27 (no. 2), 33–147.

Sutcliffe, A.J. and Zeuner, F.E. (1962). Excavations in the Torbryan Caves, Devonshire. I. Tornewton Cave. *Proc. Devon Archaeol. Explor. Soc.* 5, 127–45.

Taylor, H. (1973). The Alveston bone fissure, Gloucestershire. *Proc. speleol. Soc.* 13, 135–52.

West, R.G. (1969). Pollen analyses from interglacial deposits at Aveley and Grays, Essex. *Proc. Geol. Ass.* 62, 107–35.

West, R.G. and Wilson D.G. (1966). Cromer Forest Bed Series. *Nature, Lond.* 209, 497–8.

Wymer, J.J. (1962). Excavations at the Maglemosian sites at Thatcham, Berkshire. England. *Proc. prehist. Soc.* 28, 329–61.

7

Skeletal remains of man in the British Quaternary

THEYA MOLLESON

Abstract

Pleistocene remains of man in Britain are never abundant and are usually fragmentary. They are representative of four periods.

The skull from Swanscombe is considered to be of Middle Pleistocene age. The bones were found associated with an Acheulian industry in late interglacial deposits. The skull is morphologically advanced and probably represents an archaic *Homo sapiens*.

Homo sapiens neanderthalensis may be represented by the teeth, found associated with a cold fauna and a Mousterian industry in early glacial deposits at La Cotte de St Brelade, Jersey.

Remains of morphologically modern man come from a number of sites in central and western Britain. They fall into two groups, Middle and Late Devensian.

Introduction

The remains of Pleistocene man in Britain are few and far between, yet significant, for they represent man during important evolutionary stages from Middle Pleistocene times. Complete details will be found in Oakley, Campbell, and Molleson (1971).

Middle Pleistocene remains: Swanscombe

Perhaps the most significant remains in Britain are also the oldest. In 1935, 1936, and 1955, three fragments of the same skull were found in fluviatile deposits of the River Thames exposed in the Barnfield pit at Swanscombe, Kent, a few miles east of London. They were found at 29 m (94′) above sea-level in the oblique seam at the base of the Upper Middle Gravels, as though they had lain on the surface of the scoured-out channel of the river-bed (Oakley 1952). The gravels also yielded unworn flint hand-axes of Acheulian type together with bones and teeth of mammals representing an interglacial fauna. The remains of large mammals would seem to indicate that the Thames was bordered by woodlands interspersed by grassland. The small mammals include a number of extinct voles and *Lemmus*, presumably indicative of the onset of colder conditions (Oakley 1952). Oakley concludes that in general terms there seems to be no doubt that the Swanscombe skull dates from the closing stage of the interglacial equated with the Hoxnian.

The skull is probably that of a female because the muscular markings are slight in relation to its size and thickness. It has vertical and moderately high sides with a well-marked high nuchal line and a rounded occiput. Its cranial capacity is estimated to have been about 1325 cm^3 which is close to the average of modern female skulls. The sutural margins show no sign of closure except at one point on the sagittal suture where closure usually begins at the age of about twenty in modern Europeans.

Le Gros Clark (1938) reported that insofar as it is preserved, the skull shows no characters which would definitely exclude it from *Homo sapiens* but that it does show a number of features which would be exceptional in a modern skull. It has a broad occipital bone and the bones are exceptionally thick. In all probability the frontal bone was essentially like that of the skull from Steinheim. Thus the Swanscombe skull has come to figure as a likely ancestor for both the classic Neanderthal and Upper Palaeolithic populations.

Recently, however, Stringer (1974) has shown by multivariate analysis that Swanscombe, along with the skull from Fontéchevade, France, is most plausibly related to the early Neanderthals rather than to modern man. He concludes that in Europe we see a good record of local evolution from Heidelberg, Petralona, Vértess-zollös, Steinheim, Swanscombe, and Arago to the early and classic Neanderthals.

Early Devensian remains: La Cotte de St Brelade

The only remains in Britain attributable to *Homo sapiens neanderthalensis* are the teeth and possibly the occipital bone from La Cotte de St Brelade, Jersey. The cave is situated about 18 m above sea-level on the south coast of the island in the cliff which bounds St Brelade's bay to the eastward. It is tucked away within a large ravine or cleft which has been formed in the granite of the cliff-head.

The teeth were found during excavations by Nicolle and Sinel of the Société Jersiaise in 1910 and 1911. About 8 m of rock debris mixed with loess covered the levels of human occupation. Flint implements and flakes were plentiful near the entrance to the cave. These indicated a Mousterian occupation of the cave at a time succeeding a more temperate climate when apparently oak or elm grew in the cave.

Unfortunately the nature of the cave is such that bone is not well preserved. However, the fauna found includes remains of lemmings and voles in several thick clusters that lay uniformly on or near the top of the human deposit, notably along the eastern wall. Zeuner concluded (1940) that these deposits could not be older than Riss nor later than Würm. McBurney (1972) considers that there is evidence for a raised strandline at 18 m which would correspond to Emiliani's stage 5 and warm isotope temperatures. As the cold increased, loess deposition set in and low sea-levels allowed Mousterian hunters to make the shelter a regular port of call.

There is evidence for several levels of human occupation. The main hearth was found at 21·40m above sea-level. It was on the left-hand side of the entrance about 2 m from the opening. Just beyond and at a slightly higher level (22·35 m Burdo, 22·16 m McBurney) nine human teeth were found in clay on a rock ledge of the west wall: 4 more teeth were later found in the same area. They almost certainly belong to one individual. Keith and Knowles (1912) ascribed them to an adult, possibly female, 20–30 years of age. The teeth are in an excellent state of preservation, only the terminal parts of the roots being broken away. The crowns are smaller than those of the Heidelberg mandible: the roots, however, are in most cases, greater. The molars have fused (taurodont) roots, large neck, and a relatively small crown.

X-rays showed that secondary dentine has almost filled the pulp cavities of the molars and to a lesser degree the other teeth. This is surprising when the slight degree of wear is taken into consideration. In none of the teeth was the dentine completely exposed; only the apices of the cusps were worn away exposing limited areas of dentine.

Keith was able to use the contact facets on the proximal and distal sides of the crowns to reconstruct the shape of the dental arcade. They clearly indicated that in the upper jaw the teeth were set on a wide palate shaped, as in the Gibraltar cranium, in the form of a horseshoe, the diameter between the third molars being decidedly less than that between the second molars. The first premolar is larger than the second, a condition not usual in modern man, and its root is highly specialized showing signs of a trifid composition. The upper third molars are reduced.

Left				Right			
M_3 M_2 P_4			I_1 (r)	M_1	M_3	width at M_2	$= 68$ mm
	M_2 P_4 $P_3 C$			I_2	M_2 M_3	length of palate	$= 50$ mm

The occipital fragment was found in July 1915 by Daghorn in the ravine outside the cave about 2m beyond the entrance and nearly 6m above the level of the cave floor but under a thick deposit of yellow loess. No artifacts or fauna were found in direct association.

The fragment is from a small skull, is undistorted and shows traces of a lambdoid suture which is still open. The diploe is extremely thin as in juvenile skulls. Angel and Coon (1954) estimated the age to be about 5 years. They noted a number of features found in Neanderthal skulls but not necessarily exclusive to them.

Middle Devensian remains

During the Middle Devensian the environment in Britain was subarctic to arctic with a high tundra flora and cold fauna. The sea-levels were considerably lower than at present, allowing a dry-land connection with the European continent.

During this time Kent's Cavern was apparently an important 'base camp' for the earlier Upper Palaeolithic exploitation of Devonshire (Campbell 1972). It lies in the lower northern slope of Lincombe Hill at Torquay and consists of a large series of passages and chambers which have been formed by solution in the bedrock of Devonian limestone and which have been partly filled with deposits at various times since the Middle Pleistocene or earlier.

A human maxilla fragment was found by Pengelly in 1867 in a part of the cavern termed the 'vestibule'. The bone was deeply embedded in stalagmite twenty inches thick (Pengelly 1884). Immediately below the granular stalagmite lay the cave earth with flakes and bone implements of a mixed Upper Palaeolithic industry and a cold fauna including woolly rhinoceros and reindeer. In this layer, at 3·2 m, another right maxillary fragment was found in 1925.

Two radiocarbon dates on the earlier Upper Palaeolithic levels average out at 28 440 ± 443 BP (GrN–6201 and 6202) (Campbell 1972).

The teeth of the maxillae are very worn and large — individually as large as some of the upper teeth from La Cotte de St. Brelade — but they are not taurodont nor is the palate as wide (Duckworth 1913). In general Keith thought the maxilla most like the Aveline's Hole jaws.

Other human relics which belong to the earlier Upper Palaeolithic are the remains of three individuals from Badger Hole and the two human teeth from Picken's Hole, both sites in the Mendips. They are of entirely modern aspect.

A single canine, now lost, from Mother Grundy's Parlour, Creswell Crags, Derbyshire, marks the northern limits of the remains of man from Middle Devensian times.

Paviland Cave or Goat's Hole is a limestone cave on the north coast of the Severn estuary, Rhossili, Gower Peninsula. It is a narrow cleft, 20 m deep, in the face of the Carboniferous Limestone cliff overlooking the Bristol Channel at a height of about 10 m above present high-water mark.

The cave was first excavated by John Traherne, L.W. Dillwyn, and Miss J. Talbot between December 1822 and January 1823. Dean William Buckland joined them on 18 January (North 1942). They found many animal bone fragments in a disturbed context and mixed with recent bones and shells. Then Buckland found part of a human skeleton *in situ*.

The skeleton was beneath a shallow covering of earth (15 cm). The skull, vertebrae, and extremities of the right side were missing, probably washed away by the sea. The bones are comparatively slender but with well marked muscular impressions at least on the ulna, femur, tibia, and pelvis. The articular heads of the long bones are large and indicative of a male (Knowles *in* Sollas 1913). All the epiphyses are fused but the line of demarcation is still evident and the age is therefore not much over 25 years.

The maximum length of the femur is 476 mm and the tibia is 398 mm which give an estimated stature of 1·732 m (5′10″) according to Pearson's formula. This is similar to estimates of stature for Upper Palaeolithic hominids from France including Grimaldi, Cavillon, and Cro-Magnon. If the stature estimate is based on the length of the humerus (338 mm) it is only 1·696 m. Sollas argues that in

Table 7.1.
Remains of hominids from the British Pleistocene

Site	Found	Hominids	Stratigraphy	Culture	Date
Swanscombe, Kent	1935, 36, 55	3 skull bones	Late Hoxnian interglacial	Acheulian	(?120 000)
Pontnewydd, Denbighshire	before 1874	1 tooth	?Ipswichian	Mousterian cf. Acheulian tradn.	>47 000 (GrN-2649)
La Cotte de St Brelade, Jersey	1910–1911 1915	teeth occipital (juv.)	Early Devensian Devensian	Mousterian	>18 000 BP (BM-497)
Badger Hole, Somerset	1939 1945	juv. mandibles 2 skulls	Middle Devensian	Earlier Upper Palaeolithic	28440 ± 443 BP (GrN-6201-2)
Kent's Cavern Devonshire	1926	maxilla	Middle Devensian	Earlier Upper Palaeolithic	—
Mother Grundy's Parlour, Derbyshire	1924	canine	Middle Devensian	Earlier Upper Palaeolithic	
Paviland, Glamorganshire	1823 1912	skeleton, post-cranial bones	Middle Devensian	Earlier Upper Palaeolithic	18 460 ± 340 BP (BM-374)
Picken's Hole, Somerset	1961	teeth	Middle Devensian	?	$34\,265 \pm {}^{2600}_{1950}$ (BM-654)
Aveline's Hole, Somerset	1797 1823–1840	c. 50 skeletons, skull, jaw.	Early Flandrian	Later Upper Palaeolithic	8100 ± 50 BP (GrN-5393)
Flint Jack's Cave, Somerset	c 1893	2 skulls	Early Flandrian	Later Upper Palaeolithic	(c. 10 000)
Gough's Cave, Somerset	1903 1927-8 1928-9 1950	skeleton, frgts 4+ individuals mandible, skull frgts	Early Flandrian ?Early Flandrian ?Early Flandrian ?Early Flandrian	Later Upper Palaeolithic Later Upper Palaeolithic	9080 ± 150 BP (BM-525) (c. 10 000) (c. 10 000)

Site	Hominids	Found	Stratigraphy	Culture	Date
Kent's Cavern, Devonshire	maxilla, humerus, skull frgts	1867 1878 1925	Late Devensian or Early Flandrian	Later Upper Palaeolithic	(c. 10 000)
Langwith Cave, Derbyshire	frgts 2 skulls	1909	Late Devensian	Later Upper Palaeolithic	—
Mother Grundy's Parlour, Derbyshire	frgts 3+ skeletons	1876	Late Devensian	Later Upper Palaeolithic	—
Pin Hole, Derbyshire	frgts adult & juvenile	1925–1938	Late Devensian	Later Upper Palaeolithic	—
Robin Hood's Cave, Derbyshire	skull frgt	1969	Late Devensian	Later Upper Palaeolithic	10 390 ± 90 BP (BM-603)
Sun Hole, Somerset	tooth, arm-bone	1926–8	Late Devensian	Later Upper Palaeolithic	12 378 ± 150 BP (BM-524)
Tornewton, Devonshire	tooth		Late Devensian	Later Upper Palaeolithic	—

Upper Palaeolithic skeletons the humerus is unusually short and that Paviland man is part of the same race.

In 1912 Sollas excavated a right metatarsal and the distal end of a left humerus, neither of which belongs to the first skeleton.

The carbon-14 date obtained on bones from the skeleton was 18 460 ± 340 BP (BM-374). It pinpoints the burial to the time at or just before the Devensian glacial maximum when the glacier ice may have been only 6 km north of Paviland (Bowen 1970).

Late Devensian remains

It would appear that Britain was not exploited by man during the glacial maximum for at least 2000 years and possibly 5000 years (Campbell 1972), but after that we find strong signs of his reappearance. The total number of individuals represented by the later Upper Palaeolithic remains is about 105 against that of the earlier Upper Palaeolithic which is only 7 (Table 7. 1).

Kent's Cavern, to judge from the number of artifacts found there, can be assumed to be a 'base camp' exploiting not only the exposed English Channel plain to the east but also the potential mammal migration routes along the valleys of the Teign and Dart to the west (Campbell 1972). A skull was found outside the cave by Powe during digging in 1925. Apparently it had been laid in a fissure of the rock. The fragments of the skull were covered with a fine wash of red cave-earth. There was no associated fauna or industry.

On reconstruction the skull was seen to be rounded and high vaulted (brachycephalic). The skull is that of a young woman but already the first molars are very worn. Despite the missing occipital, Keith estimated the skull length to have been only 175 mm and the cranial capacity to be about 1400 cm^3. The face, which is present, was remarkably short and the nose snubbed (Keith 1926). Keith concludes with a picture of a woman with a wide and prominent forehead drawn in and flattened at the root of the nose, small-faced with a flat snub childlike nose.

The base of the skull is broken away and part of the palate is missing as well as the bridge and upper part of the nose. But enough remains to show that she was of the same breed as the late Palaeolithic round-headed people met with in Aveline's Hole.

Aveline's Hole, Mendips, was discovered as long ago as 1794. Probably about 50 human skeletons (now lost) were removed from the cave and on cultural criteria were attributed to the late Palaeolithic. Numerous other remains have been found from time to time, some bones have been dated by radiocarbon measurement to 9144 ± 100 BP (BM-471).

'Cheddar Man', the almost complete skeleton of an adult male, was found in Gough's Cave, Cheddar Gorge, in 1903 with no precisely recorded stratigraphic relationship to the artifacts or fauna. Dated on the bone by radiocarbon to 9080 ± 150 BP (BM-525), it must be representative of almost the last occupation of the site, in a way that recalls the burial at Paviland. It is even likely that the Upper Pleistocene people did not bury their dead in caves that were being actively occupied but rather in formerly important sites that may or may not have had some ritual significance.

Fragmentary remains of later Upper Palaeolithic man have been found at Sun Hole, Mendip Hills, and possibly at Flint Jack's Cave, although the provenance for these last is doubted.

In the Creswell Crags area, Derbyshire, four caves have yielded hominid remains though none represents a burial. A hominid frontal bone was found by John Campbell during excavations in 1969 in Robin Hood's Cave and from both Pin Hole Cave and Langwith Cave there are fragments of a juvenile and adult.

From the paucity of hominid remains in Britain it is apparent even during interglacial times that it is unlikely that the Palaeolithic population of Britain was ever very great. So long as there was the possibility of free movement between France and Britain these populations would probably not have been genetically isolated. It is probable that the hand-axe people as at Swanscombe were mainly adapted to life in open grassland rather than to wooded country (Oakley 1952). Later peoples were hunter–gatherers with a strong preference for caves for shelter. They may have migrated out of Britain during the winters (Campbell 1972).

References

A more complete bibliography will be found in Oakley, K.P., Campbell, B.G., and Molleson, T.I. (1971). *Catalogue of Fossil Hominids, Part II: Europe*, pp. 17-43. British Museum (Natural History), London.

Angel, J.L. and Coon, C.S. (1954). La Cotte du St Brelade II: Present Status, *Man*, **54**, 53-5.

Bowen, D.Q. (1970). The Palaeoenvironment of the 'Red Lady' of Paviland, *Antiquity* **44**, 134-6.

Campbell, J.B. (1972). The Upper Palaeolithic of Britain. Thesis, University of Oxford.

Le Gros Clark, W.E. (1938). General features of the Swanscombe skull bones, *Jl. R. anthrop. Inst.* **68**, 58-61.

Duckworth, W.L.H. (1913). Notes on some points connected with the excavation of Kent's Cavern, Torquay, *J. Torquay nat. Hist. Soc.* **1**, 215-20.

Keith, A. (1926). Report on a human skull found near the North entrance to Kent's Cavern, *Trans. Proc. Torquay nat. Hist. Soc.* **4**, 289-94.

Keith, A. and Knowles, F.H.S. (1912). A description of teeth of Palaeolithic Man from Jersey. *J. Anat. Physiol. Lond.* **46**, 12-27.

McBurney, C.B.M. (1972). The Cambridge excavations at La Cotte de St Brelade Jersey—a preliminary report. *Proc. prehist. Soc.* **37**, 167-207.

North, F.J. (1942). Paviland Cave, the 'Red Lady', the deluge and William Buckland. *Ann. Sci.* **5**, 91-128.

Oakley, K.P. (1952). Swanscombe Man. *Proc. Geol. Ass.* **63**, 271-300.

Pengelly, W. (1884). The literature of Kent's Cavern. Part V. *Rep. Trans. Devon. Ass. Advmt. Sci.* **14**, 189-434.

Sollas, W.J. (1913). Paviland Cave: An Aurignacian Station in Wales, *Jl. anthrop. Inst.* **43**, 1-50.

Stringer, C. (1974). Population relationships of Later Pleistocene hominids: A multivariate study of available crania, *Jl. Archaeol. Sci.* **1**, 317-42.

Zeuner, F.E. (1940). The Age of Neanderthal Man with notes on the Cotte de St Brelade, C.I. *Occ. Pap. hist. Archaeol. Univ. Lond.* **3**, 10-13.

8

The archaeology of man in the British Quaternary

Abstract

The sequence of Palaeolithic industries in Britain is considered against the framework of the stratigraphical stages currently employed for the British Quaternary. Four species of industries are recognized: Clactonian, Acheulian, Levalloisian, and Mousterian. Lower Palaeolithic activity is most evident from the Hoxnian to the Early Devensian stages. The earliest artifacts are probably of Cromerian age. On present evidence, all of the British Lower Palaeolithic is considerably more recent than its African counterparts. There were probably two main phases of the Upper Palaeolithic, both in the Late Devensian, but separated by the harshness of glacial conditions near its close. The late Upper Palaeolithic is considered to have been a strong influence upon the succeeding Mesolithic period. Some of the key sites are shown on a generalized time chart.

The Lower Palaeolithic

The archaeological evidence for the existence and attainments of man in Britain during the Quaternary period has to be interpreted against two particular realizations: the generally accepted one that for much the greater part of this time Britain was not isolated from the continent, and the ever-increasing volume of new evidence that man has existed outside Europe for considerably longer than he has within it. Dating is therefore of crucial importance for relating the cultural stages of man in Britain against his achievements elsewhere, and it is clear that cultural stages can only be assessed in a very limited manner by a study of flint implements alone. Associated material is required to enable at least some tentative reconstruction of the contemporary environment and human activity. At the same time it is necessary to place discoveries within a temporal sequence, especially as the evolution of human societies may rightly be regarded as the prime concern of the archaeologist.

The influence of the late Professor F.E. Zeuner did much to bring this about, for he guided the then-prevalent concentration on archaeological sequences based on the flimsy basis of the typology of derived artifacts, into what he termed 'geochronology', coupled with environmental archaeology (Zeuner 1952). It is irrelevant that his acceptance of the Milankovitch Time-Scale and his interpretation of the events in the British Quaternary against it render many of his conclusions unacceptable in 1977, for the advances in knowledge which make them unacceptable have mainly been stimulated by application of the methods he proposed. Archaeologists have concentrated within the last decade or so on two complementary aspects for the Lower Palaeolithic period; collating and describing the evidence on a non-selective or objective basis, and attempting to locate sites where they could apply the archaeological techniques normally employed in the excavation of later human periods. The former has caused the publication of the Council for British Archaeology's Gazetteer of British Lower and Middle Palaeolithic Sites (Roe 1968) and a more detailed gazetteer of one area, the Thames Valley (Wymer 1968). Hand-axe groups have been identified by series of objective measurements (Roe 1964, 1969). Major excavations which fulfil the latter requirement have been conducted at High Lodge, Mildenhall, Suffolk by Sieveking in 1964–5; at Clacton-on-Sea, Essex in 1969–70 (Singer *et al.* 1973); Swanscombe, Kent in 1968–71 (Waechter *et al* 1969, 1970, 1971, 1972); and at Hoxne,† Suffolk

†Supported by the United States National Science Foundation Grants no. GS 2907 and GS 41435 (to Dr. R. Singer, University of Chicago).

(Wymer 1974). The result of such work, in relation to results achieved by other Quaternary disciplines, will be considered first, before assessing the Upper Palaeolithic and Mesolithic periods in Britain.

First, there is the question of the identity and taxonomy of so-called industrial assemblages in the Lower Palaeolithic. British archaeologists would all agree that there exist three distinct series:

Clactonian—industries dominated by chopper-cores and flakes with un-specialized flake tools;
Acheulian—industries dominated by hand-axes of various forms and associated with non-specialized and specialized flake tools;
Levalloisian—industries dominated by flakes struck from prepared cores, sometimes associated with hand-axes, otherwise regarded as a technique rather than an industry.

To these can be added 'Mousterian', identified on the grounds of typological parallels with the continental Mousterian, combined with stratigraphical, faunal, or other information that suggests a date during the last glaciation.

Typological variations within each series are numerous, but most workers would now consider it unwise to accept any sequence based solely on the supposition that technological refinement or development coincided with temporal succession. Roe's objective statement on the existence of hand-axe groups in Britain carefully avoids any suggestion that a sequence is implied. However, there is still a tendency for archaeologists to regard Acheulian industries dominated by hand-axes displaying crude technique to be earlier than more refined industries, even if the 'industry' is the result of uncontrolled discovery, less than a dozen artifacts, or even a single specimen (Roe 1975). It is difficult to resist this view, but it is necessary to repress it until suitable dating can be applied to the artifacts. Likewise, the Clactonian Industry clearly represents a low level of technology, but it would be unwise to suggest it belongs at any particular place in the Palaeolithic sequence just because of this. There is no reason why we should not expect to find in Britain highly refined and developed flint industries at the very beginning of the sequence, especially if it can be demonstrated that they already existed in other parts of the world. The African evidence suggests very strongly that this could be the case.

Dating, at least in the relative sense, is therefore of paramount importance in constructing any sequence upon which we might obtain some idea of what was happening in Britian during the Palaeolithic period. Dating is therefore the next consideration, before discussing what the industries may mean in terms of human social, economic, or even physical evolution.

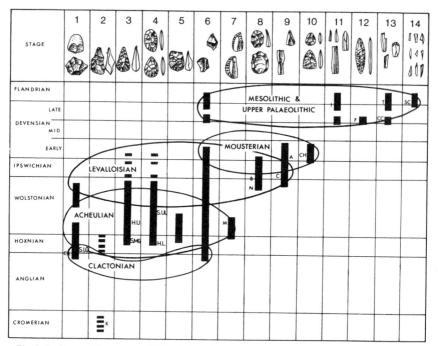

Fig. 8.1. General Sequence of flint industries in the British Quaternary. Main tool types (1–14) are featured with their temporal range, to give a simplified visual statement of the assemblages comprising the five broad divisions of the Palaeolithic and Mesolithic periods. The more problematical or indefinite parts of the sequence are shown as broken columns. The estimated stratigraphical position of a few key sites are indicated against the vertical columns by letters.

Tool Types

1. Pebble and bi-conical chopper-cores. Characteristic of the Clactonian Industry, as at Clacton-on-Sea and Swanscombe (Lower Gravel and Lower Loam). Such primitive arti-facts are also found as a minor component of Acheulian and proto-Levalloisian Indus-tries. Bi-conical cores were common in the Hoxne Upper Industry, presumably for pro-ducing the flakes from which the specialized flake tools were made.

2. Crude, stone-struck hand-axes. These occur in high level gravels at Fordwich, Kent, Farnham, Surrey, and Corfe Mullen, Dorset, but the dating evidence and associations are unsatisfactory. There are also several hand-axes of this type from West Drayton and Yiewsley, in the Thames Valley. A series from Kent's Cavern, Devon, may be associated with a Cromerian fauna, and is the justification for the inclusion at this stage on the table. Artifacts, with a certain Cromerian or intra-Anglian association, are known from Westbury-sub-Mendip, Somerset, but no definite tool types have yet been identified.

3. Pointed hand-axes, generally finished with shallow-flaking at the thin end. Typical of the Middle Gravels at Swanscombe, associated with the Swanscombe skull. Also found in the Hoxne Upper Industry and a likely or known component of most Acheulian Industries.

4. Ovate and cordate hand-axes, often of fine workmanship, sometimes sharpened with tranchet blows and sometimes with twisted edges. Typical of the Hoxne Lower Industry and also the Swanscombe Upper Loam Industry. Also found in Levalloisian contexts at Baker's Hole, Northfleet, Kent, and at Crayford, Kent.

5. Cleavers. Rare in the British Palaeolithic sequence, but found in the Thames Valley more frequently than elsewhere, mixed with undifferentiated Acheulian assemblages.

6. Flakes with unspecialized secondary working. Ubiquitous tools found in all stone industries, but particularly characteristic of the Clactonian.

7. Heavy, specialized flake tools, usually classified as side-scrapers and variants thereof. Often found in undifferentiated Acheulian assemblages, but absent in the Swanscombe Middle Gravels and the Hoxne Lower Industry. They occur in the Hoxne Upper Indus-

try and the Swanscombe–Northfleet Upper Loam. The finest series is from High Lodge, Mildenhall.

8. Prepared cores ('Tortoise' cores) and the flakes finally removed from them. The striking platform generally bears distinct traces of the final preparation of the core edge, to ensure the correct striking angle for the successful removal of the flint flake across the core's upper face. Found in large numbers on 'factory' sites such as Baker's Hole, North-fleet. Also found in late Wolstonian–Ipswichian contexts at Brundon, Suffolk, and above terrace gravels at West Drayton and Iver in the Thames Valley.

9. Flake-blades and pointed flake-blades with prepared striking platforms. These are removed by Levalloisian technique and are generally in association with the previous category. They represent a specialized development of radially-struck tortoise cores, as these flakes have been removed from converging or near-prismatic cores. They occur at Baker's Hole, Northfleet, and dominate the apparently later industries at Crayford, Kent and Acton, London.

10. Small cordate and flat-butted cordate (*bout coupé*) hand-axes. Typical of the early part of the Devensian and thus found in low-lying terrace gravels or buried channels of existing rivers, e.g. at Ipswich in the Gipping Valley, at Little Paxton, St. Neots, Hunting-donshire, and at Christchurch, Hampshire. Also dated by discoveries in cave sites at Coy-gan Cave, Carmarthenshire, Oldbury, Kent and Kent's Cavern, Devon.

11. Blades, struck from single or double-platform prismatic cores. They are characteristic of the Upper Palaeolithic and Mesolithic periods, although a few may occur in earlier developed Levalloisian Industries.

12. Leaf points, worked either unifacially, bifacially or part-bifacially. Found in a few well-dated early Upper Palaeolithic cave sites (Ffynnon Beuno, North Wales, Robin Hood's Cave, Creswell Crags) and also in late Devensian gravels of the River Gipping near Ipswich, Suffolk.

13. Small, specialized flake tools, including gravers, scrapers, borers, backed blades, and numerous other categories. Typical of the Upper Palaeolithic and Mesolithic Industries.

14. Microliths. Although a few microliths are found in late Upper Palaeolithic contexts, they are virtually diagnostic of the Mesolithic. Later Mesolithic microliths tend to be smaller and of rod-like and geometric shapes. The hafted tranchet axe (not depicted) is also characteristic of the Mesolithic period, and does not occur in the Upper Palaeolithic period.

Bone Tools

There is no certain evidence for the use of bone for tools and weapons in the British Lower Palaeolithic. They first occur in the early Upper Palaeolitic, as at Kent's Cavern, some of the Cheddar Caves, and at Creswell Crags (mainly pins and awls). Barbed points of bone and antler occur in the late Upper Palaeolithic (e.g. at Kent's Cavern and Ave-line's Hole, Somerset) and uniserial barbed points were typical of the early Mesolithic site at Star Carr, Yorkshire, where groove-and-splinter work was practised. Bone and antler were also used for tools and weapons at the early Mesolithic site at Thatcham, Berkshire, but no barbed points have been found there, or trace of groove-and-splinter work.

Key to Sites

A.	Creffield Road, Acton, Ealing London Borough
B.	Brundon, Sudbury, Suffolk
C.	Crayford, Kent
CC.	Robin Hood's and Pinhole Caves, Creswell Crags, Derbyshire
CH.	Christchurch, Hampshire
CL.	Clacton-on-Sea, Essex
F.	Ffynnon Beuno, Flintshire
H.L.	Hoxne, Suffolk (Lower Industry)
H.U.	Hoxne, Suffolk (Upper Industry)
I.	Sproughton, Ipswich, Suffolk
K.	Kent's Cavern, Torquay, Devonshire
M.	High Lodge, Mildenhall, Suffolk
N.	Baker's Hole, Northfleet, Kent
S.LG.	Swanscombe, Kent (Industry in Lower gravel and Lower Loam)
S.MG.	Swanscombe, Kent (Industry in Lower and Upper Middle Gravel)
S.UL.	Swanscombe, Kent (Industry in Upper Loam)
SC.	Star Carr, Yorkshire
T.	Thatcham, Newbury, Berkshire

There is sufficient information from Clacton-on-Sea and Swanscombe to state categorically that the Clactonian Industry *at those two sites* is something separate and distinct from the other series (Acheulian and Levalloisian). It is the very lack of hand-axes, specialized tools or use of Levalloisian technique that defines it. Whether the people responsible for the industry at those two sites made hand-axes elsewhere is irrelevant, as it is a speculation beyond any possible proof. There are grounds for considering the industry at Clacton, and probably Swanscombe, as the earliest 'identifiable' industry in Britain, although they are not the earliest artifacts (see below). It would appear to be earlier than the majority, or perhaps all, of the Acheulian industries. Clacton offers the better possibilities for dating, although at this stage of Pleistocene studies, with no suitable radioactive methods of obtaining absolute or chronometric dates, it must be admitted that little more can be done than attempt to insert the site (as with all other Palaeolithic sites) into a framework of stages based on climatic oscillations; stages recognized and related to each other with varying degrees of confidence through stratigraphy, pollen analysis, and faunal analysis. The archaeologist is, certainly for the Middle Pleistocene, totally dependent on the frameworks established by his colleagues, the most recent expressions of which are to be found in two publications of the Geographical Society (Mitchell *et al.* 1973; Evans 1971).

For Clacton, Gladfelter has argued (in Singer *et al.* 1973) that the presence of Greensand chert and bi-zoned flint suggests a southerly source for the gravel containing the industry, which would have been non-existent after the diversion of the Thames into its present valley system, the Stage III deposits of Wooldridge and Linton (1955). This diversion could have effectively barred the northward-flowing Medway and allied river systems from transporting chert and bi-zoned flint much beyond the present line of the Thames estuary. Such implies a date prior to the maximum extension of the ice sheet relegated to the Anglian Stage. However, the Clacton Channel is cut through gravel (the Holland Gravel) which contains minerals derived from the Chalky Boulder Clay, which is presumably the till of the Anglian Glaciation. Thus, the Clacton gravels on different lines of evidence are both earlier and more recent than the Anglian Ice! Pollen analyses by Mullenders (Wymer 1974) in marl overlying the gravel have been interpreted by him as indicative of a Pre-temperate interglacial zone. This has been questioned by West (personal communication); the most likely, safest conclusion is to place the industry within the dating bracket of Late Anglian–Early Hoxnian. This fits in suitably with Turner's dating of the freshwater beds at Clacton itself, i.e. 2 km distant from the town, as the Early-temperate zone of the Hoxnian Interglacial (Turner and Kerney 1971).

At Swanscombe, the molluscan and faunal evidence alone is sufficient to justify a correlation with Clacton, both in the Lower Gravel and Lower Loam. Mullenders, on the basis of pollen analyses, would equate the bottom of the Lower Loam at Swanscombe with its Clactonian material in primary context, as the Early-temperate zone of the Hoxnian.

At Barnham, Suffolk, a Clactonian industry is in gravel (Paterson 1937) which may be outwash of the 'Lowestoft Glaciation', i.e. the Anglian. Thus, although there are several points of conflict, there is reasonable evidence to place the Clactonian industries from these sites within a time-range spanning the later stages of the Anglian to the Early-temperate zone of the Hoxnian. There are no other sites in Britain (excluding Little Thurrock, Grays, and Rickson's Pit, Swanscombe, which is presumably an extension of the Swanscombe Lower Gravel) where a Clactonian Industry has been found in an unmixed or primary context. Where such an industry has been identified in the Thames Valley, the only grounds for separating it from hand-axes in the same deposit are those of differing conditions of the artifacts. This is not satisfactory, and all that may be said with any confidence is that none of the gravels containing the apparent Clactonian Industry is likely to have been deposited prior to the Hoxnian Interglacial, or after the Wolstonian Stage.

The pollen evidence from Hoxne shows that the earliest Acheulian Industry represented at that site is in the latter part of the Early-temperate zone, thus more recent than that at Clacton. At Swanscombe, the Clactonian is overlain by gravels with Acheulian tools so, in both cases, Clactonian precedes Acheulian, and this leads to the intriguing questions: when was man first in Britain and what manner of flint industry did he employ? There have been many claims for very early artifacts, particularly in the crags of East Anglia. Most archaeologists hesitate to accept the evidence from the crags, mainly because so much that has been claimed has been on a misunderstanding of the elementary principles of flint-knapping. However, there are several flakes from the crags which, in other contexts, would be regarded as human workmanship. It is a matter of opinion, and until some satisfactory assemblage occurs must remain unproven. It was thus a most significant and fortunate discovery when a few flint artefacts were found in Somerset in 1974, in unequivocal association with a mammalian fauna that is broadly Cromerian, and certainly pre-Anglian in date (Bishop 1975). Unfortunately the number of flint artifacts at this site, Westbury-sub-Mendip, is very small and it is impossible to define any industry. There is only one piece that is more than a flake, and it could be regarded equally as a small, crude, broken hand-axe or a polyhedral core. The writer is more of the opinion that it is the former. There is also a unique series of crude, stone-struck hand-axes from Kent's Cavern which may have been associated with a Cromerian fauna. Claims have been made for early Acheulian, in a temporal and technological sense, in high level gravels at various places (Fordwich, Kent; the Caversham Channel; Farnham, Surrey), especially by Roe (1969), but it does not seem possible to substantiate them. The gravels in question seem unlikely to date prior to the Anglian, although the artifacts could of course have been derived from much earlier surfaces or sediments. It is tempting to make correlations with better-documented sequences in Africa and archaeologists may perhaps be forgiven for sometimes wondering whether the great hiatus between the beginnings of human industrial activity in Africa and the first evidence for man in Britain is really as great as it would at present appear to be.

It could be argued that there was considerable human activity in Britain prior to the Anglian Glaciation, but the evidence for it was destroyed by the very passage of the ice and the effects of its out-wash, and all that remains are the derived artifacts in the post-Anglian gravels. The writer has been puzzled for many years by the uncanny similarity in the distribution of Wolstonian till and the sparsity of Lower Palaeolithic material in the same area. The explanation could be that the Wolstonian ice sheet again effectively destroyed the evidence for human occupation during the preceding interglacial over nearly all the area it covered. It seems inconceivable that once man was in Britain, he confined his activities to southern England, and it could also hardly be expected that any archaeological material he left behind would remain in a primary context after at least two glaciations, or even be recognizable. Undisputed artifacts in gravels beneath tills in Britain are, as may be expected, extremely rare, and the only site which would seem to have a clear stratigraphical re-lationship is one at Welton-le-Wold, Lincolnshire (Alabaster and Straw 1976) where the overlying till is considered to be Wolstonian. Of greater import is the lack of all archaeological material in the Thames Valley in gravels which would appear to pre-date the Stage III diver-sion of the Thames into its present course south of the boulder clay. This absence cannot be explained away by insufficient exposures, as may be the case elsewhere, for the Winter Hill Gravels of the Thames Valley, dug commercially over large areas in the Uxbridge district, have never yielded a single artifact. These gravels have been dated by Wooldridge to what may now be termed the Anglian Stage, on the strength of their being overlain by chalky boulder clay in the Vale of St. Albans.

The quantity of Acheulian material in sediments, mainly gravels, that are post-Anglian till is prodigious in south east England, particularly south east of a line drawn from the Wash to the Solent, i.e. beyond the line of the main advance of the Wolstonian ice-sheet. The majority are in a derived condition, but are unlikely to have travelled very far, and considerable human activity is implied within this area although, as stated above, the Wolstonian ice may have obliterated the evidence for occupation in the rest of Britain. It is also significant that Acheulian implements are rare in gravels that post-date the Wolstonian Stage. If the Taplow Terrace gravels of the Thames are late Wolstonian, and not Devensian, then it could be said that there was little Acheulian in Britain by that time. The evidence from Hoxne is sufficient to show that there were Acheulian hunting groups visiting the Hoxne lakeside during the middle and latter part of the Hoxnian Interglacial. The industry at this stage at this site is highly refined and evolved in the technological sense. Elsewhere in Britain, it is very difficult to be sure whether the Acheulian sites, whether derived in gravels or in a primary context (such as Foxhall Road, Ipswich) are of Hoxnian or Wolstonian date. Even the Swanscombe skull, and its associated Acheulian industry of pointed hand-axes, could belong to some early phase of the Wolstonian and not the Late Hoxnian as suggested by Kerney (1971).

Two important sites where there are complex series of sediments associated with flint industries, both probably of early Wolstonian date, are Hoxne (upper levels) and High Lodge, Mildenhall. The many sites suggest a variety of Acheulian assemblages based on differing inherited traditions or perhaps local circumstances. It would seem that the shape and refinement of the hand-axes had little or nothing to do with any temporal succession. However, as a last and cautious support for typology, the large, crude stone-struck hand-axes which might belong to earlier Acheulian Industries *are* lacking, at both Hoxne and Swanscombe. The hand-axes of these various Acheulian industries are accompanied sometimes by non-specialized worked flakes, and sometimes by fine, elegant scrapers. Bi-conical cores suggest occasional knapping for the sake of flakes. The Hoxnian and Wolstonian Stages would certainly be the time of maximum Lower Palaeolithic occupation of the country; the Hoxnian (by definition and demonstration) is a clear interglacial stage, passing through a smooth succession from late-glacial and temperate to early-glacial zones; the Wolstonian Stage includes climatic oscillations, and could well be just a useful name to cover a whole series of changes covering a long period of the Pleistocene. This is a geological problem to which the archaeologist desperately awaits a solution, for it involves the majority of his evidence for the Lower Palaeolithic in Britain.

The third series to consider is the Levalloisian, whether it be considered an industry or a technique. Although certain controlled forms of knapping in the Acheulian certainly anticipate its use, it cannot really be accepted that there is a Levalloisian element in the Acheulian Industries which appear to date to the Hoxnian or early Wolstonian, i.e. Swanscombe Middle Gravels and Hoxne. At Hoxne itself there is nothing whatsoever, although in gravels of the Waveney which post-date the Hoxne sequence and are probably late Wolstonian, derived Levallois flakes have been found. At Northfleet the classic Baker's Hole Industry clearly post-dates the whole of the Swanscombe Barnfield Pit sequence and is generally regarded as late Wolstonian. Proto-Levalloisian industries also appear to date about this time, on the evidence of a site at Purfleet.

Hand-axes sometimes accompany the Levalloisian material, as at Baker's Hole and Crayford, where a rich industry with a high proportion of blades with prepared platforms was found in an apparently early Ipswichian context (Chandler 1914). The industry at Creffield Road, Acton, contains both flake-blades and pointed flake-blades. It is a fine silt ('brick-earth') at about 25 m above the present Thames Valley floor, and may date to the last glaciation (Devensian). There is thus implied a time span from the latter part of the Wolstonian to the early part of the Devensian, with typological variations that could indicate a real technological and cultural change.

One type of hand-axe which does seem to be confined to a particular industry and time span is the flat-butted cordate hand-axe, otherwise known by the French term, *bout coupé*. In all cases where there is some associated dating evidence, the implications are that they belong to the earlier phases of the last glaciation. The negative evidence that they do not occur in earlier contexts is equally convincing. They are known from river sediments of Devensian date (e.g. Christchurch, Hampshire; Ipswich, Suffolk) and in caves (Coygan, Carmarthenshire; Oldbury, Kent; Kent's Cavern, Devonshire) and, on continental parallels, can be described as Mousterian. It is certainly the last expression of hand-axe manufacture in Britain.

Complementary to the sequence of Palaeolithic industries is the economy and environment, if not the whole ecology, of the human groups responsible for them. Only sites with material in a primary context, with sufficient associated faunal, floral or other evidence, and excavated under controlled conditions, can possibly assist with this aspect. The number of such sites in Britain that either have or could satisfy these conditions is deplorably small. At present only a little can be inferred. At all stages it would seem that the groups were dependent on hunting large game, for the association of faunal remains with flint tools at numerous sites must reflect butchering activities. Groups with a Clactonian Industry hunted deer, ox, elephant, and rhinoceros among other animals; groups with an Acheulian Industry, such as at Swanscombe, hunted similar animals, although horse was more plentiful. The Hoxne evidence suggests horse, deer, ox, and elephant were the most hunted animals during those stages of the Acheulian. It is intriguing that such dissimilar flint industries as the Clactonian and Acheulian should occur where, as far as butchering was concerned, the needs would presumably have been the same. Nothing is yet known of the vegetable content of the Palaeolithic diet, but macro-floral studies at favourable sites might eventually give some clues. There is also the inference that all groups preferred situations beside large sheets of water, either lakes or rivers. The major concentrations of palaeoliths along the Thames Valley are in the region of the confluence of important tributaries, but possibly this has been caused more by natural sorting agencies. Open sites are ubiquitous and not surprisingly, as natural caves or rock shelters are non-existent through much of the area of densest Palaeolithic finds. Yet Acheulian material has been found in Kent's Cavern in circumstances which suggest it could predate most of the British Acheulian. At the other end of the time scale, Mousterian Industries occur in caves within Devensian sediments. There is no evidence to suggest the use of caves in the Hoxnian or Wolstonian Stages.

One method of study which promises to do much to explain the past activities among Palaeolithic groups is that of critical examination of the micro-wear or edge damage on flint artifacts. Such study, coupled with the analysis of horizontal distributions of both the artifacts and any other associated remains, obviously requires material found in a primary context. Recently excavated flints from Clacton and Hoxne promise useful information of this type. Human skeletal remains for the whole of the Lower Palaeolithic period in Britain are confined to the Swanscombe skull, so the possibility of differing human physical types being associated with different industries can only be speculative.

The Upper Palaeolithic

The Upper Palaeolithic period in Britain is known mainly from cave sites. There have been few recent discoveries, but considerable review of the existing evidence and some re-excavation of significant sites, such as Creswell Crags, have taken place. John Campbell (unpublished thesis) has done much to clarify the interpretation of the not very prolific material by discarding any attempt to correlate it with the better-documented French Upper Palaeolithic. He has suggested a sub-division into British early Upper Palaeolithic and British later Upper Palaeolithic. The Early episode would appear to equate with the relatively warmer conditions of the Middle to first part of the Late Devensian, and be separated from the Later Upper Palaeolithic by the harsh glacial conditions of the Late Devensian. This would seem likely but is not proven. However, both these stages of the British Upper Palaeolithic are within the range of radiocarbon dating, and this problem should be resolved in future. A list of the most useful or reliable dates, with a valuable commentary, has been published by Mellars (1974).

The early Upper Palaeolithic is characterized by leaf points, as at Robin Hood's Cave, Ffynnon Beuno, and Kent's Cavern, and it is now thought more likely that these have affinities with the *blattspitzen* of central and eastern Europe rather than with the Solutrian of France. A few examples of these leaf points have come from river valleys in East Anglia, and also from the Thames and, if these implements really are of the same period, they constitute almost the only evidence for open sites.

The later Upper Palaeolithic is better represented and, to judge by radiocarbon dates, may have spanned four thousand years (12 000–8000 B.C.), i.e. most of the late glacial period. Much of the flintwork is typified by backed blades and generally referred to as a Creswellian Industry. Bone tools and biserial barbed points occur. Horse, reindeer, and giant deer were the main game animals. It is unlikely that occupation was ever very intense during this long period, for the sites are few and, on continental standards, small. This was a period of very low sea level, and communication with the rest of Europe may have been easy. Small bands of itinerant hunters are inferred and their more permanent camps may have been in the lowlands now covered by the North Sea or English Channel. In this respect it may not be coincidental that the two most significant open sites so far discovered, near Ipswich, Suffolk, and at Hengistbury Head, Hampshire, are both close to the present shoreline. These two sites are not Creswellian, but the former is dominated by long blades, and the latter typified by shouldered or tanged points. European connections are indicated, especially perhaps with the Low Countries and the reindeer camps of the Hamburg region. Uniserial barbed points have been found in late glacial deposits in Lancashire and Suffolk, and it is possible that some of the Yorkshire ones may be earlier than previously suspected (Wymer *et al.* 1975).

The Mesolithic

With this background of later Upper Palaeolithic activity in Britain, the formal definition of the Mesolithic period as that post-dating the late glacial (i.e. more recent than pollen Zone III Younger Dryas) seems unreal, even if convenient. The early Mesolithic industries of Britain, characterized by the dominant use of microliths in their hunting equipment, and the hafted flint axehead, can no longer be regarded as a simple incursion of Maglemosian hunters across the North Sea from Denmark before the breaching of the English Channel, especially when the British radiocarbon dates are earlier than any of the Danish ones. Indigenous developments and influences from various parts of Europe are the more likely explanations; groups with varying ingrained traditions but practising a basically similar economy. Their more specialized flint industries, found in quantity, are more susceptible to metrical and typological analysis than the more generalized equipment of the Palaeolithic, and varying traditions can be identified with greater confidence. Jacobi and Switsur (personal communication) have claimed to have separated some of the Early British Mesolithic material by computer cluster analysis, and there is a hopeful prospect for this purely archaeological approach. Complementary to such studies is the cultural and environmental evidence to be obtained from detailed, controlled horizontal excavation of living sites. Mesolithic camping sites were frequently in river valleys, close to the water's edge, and their occasional submergence beneath Flandrian sediments has preserved some of these sites with optimum conditions for archaeological investigation.

The extensive distribution of Mesolithic sites in Britain has been emphasized recently by the production of a gazetteer of Mesolithic sites in England and Wales by the Council for British Archaeology. It will be several years before this mass of material can be classified in terms of time, tradition, population movement and integration, cultural and economic status, industrial and aesthetic achievements, and suchlike, but there seems good hope that it will be done, and we shall have a clearer conception of the indigenous population that witnessed, probably with alarm and bewilderment, the arrival of the Neolithic settlers in the fourth millenium.

References

Alabaster, C. and Straw, A. (1976). The Pleistocine context of faunal remains and artefacts discovered at Welton-le-Wold, Lincolnshire. *Proc. Yorks. geol. Soc.* 41, 75-94.

Bishop, M.J. (1975). Earliest record of man's presence in Britain. *Nature, Lond.* 253, (5487), 95-7.

Chandler, R.H. (1914). The Pleistocene deposits of Crayford. *Proc. Geol. Ass.* 25, 61-70.

Evans, P. (1971). The Phanerozoic time-scale. A supplement. Part 2. Towards a Pleistocene time-scale. *Geol. Soc. Lond. Spec. Pub.* 5.

Kerney, M.P. (1971). Interglacial deposits in Barnfield Pit, Swanscombe, and their molluscan fauna. *J. Geol. Soc. Lond.* 127, 69-93.

Mellars, P.A. (1974). The Palaeolithic and Mesolithic. In *British Prehistory, a new outline* (Ed. C. Renfrew), pp. 41-99, 268-279. Duckworth, London.

Mitchell, G.F., Penny, L.F., Shotton, F.W. and West, R.G. (1973). A correlation of Quaternary deposits in the British Isles. *Geol. Soc. Lond. Spec. Rep.* 4.

Paterson, T.T. (1937). Studies in the Palaeolithic Succession in England. No. 1. The Barnham Sequence. *Proc. prehist. Soc. E.Anglia.* 3, 87-135.

Roe, D.A. (1964). The British Lower and Middle Palaeolithic: some problems, methods of study and preliminary results. *Proc. prehist. Soc.* 30, 245-67.

Roe, D.A. (1968). A Gazetteer of British Lower and Middle Palaeolithic sites. *Research Rep. Council for British Archaeology,* 8, 1-356.

Roe, D.A. (1969). British Lower and Middle Palaeolithic hand-axe groups. *Proc. prehist. Soc.* 34, 1-82.

Roe, D.A. (1975). Some Hampshire and Dorset hand-axes and the question of Early Acheulian in Britain. *Proc. prehist. Soc.* 41, 1-9.

Singer, R., Wymer, J.J., Gladfelter, B.G. and Wolff, R. (1973). Excavation of the Clactonian Industry at the Golf Course, Clacton-on-Sea, Essex. *Proc. prehist. Soc.* 39, 6-74.

Turner, C. and Kerney, M.P. (1971). The Age of the freshwater beds of the Clacton Channel. *J. Geol. Soc. Lond.* 127, 93-5.

Waechter, J. d'A. (1972). Swanscombe 1971. *Proc. R. anthropol. Inst. 1971* 73-78.

Waechter, J. d'A. and Conway, B.W. (1969). Swanscombe 1968 (interim report on the excavations in the Barnfield Pit) *Proc. R. anthropol. Inst. Lond. 1968,* 53-61.

Waechter, J. d'A., Newcomer, M.H. and Conway, B.W. (1970). Swanscombe 1969 (Barnfield Pit, Kent). *Proc. R. anthropol. Inst. Lond. 1969,* 83-93.

Waechter, J. d'A., Newcomer, M.H. and Conway, B.W. (1971). Swanscombe 1970. *Proc. R. anthropol. Inst. 1970,* 43-64.

Wooldridge, S.W. and Linton, D.L. (1955). *Structure surface and drainage in South-East England.* George Philip, London.

Wymer, J.J. (1968). *Lower Palaeolithic Archaeology in Britain, as represented by the Thames Valley.* John Baker, London.

Wymer, J.J. (1974). Clactonian and Acheulian industries in Britain. Their chronology and significance. *Proc. Geol. Ass.* 85, (3), 391-421.

Wymer, J.J., Jacobi, R.M. and Rose, J. (1975). Late Devensian and Early Flandrian barbed points from Sproughton, Suffolk. *Proc. prehist. Soc.* 41, 235-41.

Zeuner, F.E. (1952). *Dating the past.* (3rd Edn.) Methuen, London.

9

Quaternary history of the British flora

SIR HARRY GODWIN

Abstract

By the use of the Cambridge Data Bank of Quaternary plant records, now with entries exceeding 50 000 and including more than half of the present British Flora, it has been possible to recognize in each of the four latest interglacials, Cromerian, Hoxnian, Ipswichian, and Flandrian, floristic-vegetation changes corresponding to the climatic cycle of pre-temperate, early-temperate, late-temporate, and post-temperate. With the broad repetition of the vegetational cycle there has been progressive loss of Tertiary elements.

There is abundant evidence to allow reconstruction of the flora and vegetation during the last (Devensian) cold stage and to recognize how important were the continentality of climate, widespread salinity, and extreme diversity of topography and soil. The treelessness was broken during some interstadia (Chelford) but not during others (Upton Warren). Despite scantier records, earlier glacial stages present similar patterns of vegetational behaviour and to a large extent the same species.

Flandrian vegetational history, based on very abundant fossil evidence, indicates that we are now in a late temperate interglacial stage and allows us to follow patterns of persistence through, and re-invasion after, the Devensian, as well as post-Neolithic human destruction of natural vegetation.

Evidence of plant remains

It is now generally accepted that the history of the British Flora has to be based, not upon conjecture as to the significance of the distribution patterns of our present native plants, but upon the factual evidence of Quaternary geological records and increased knowledge of Quaternary events and environments. Macrofossil remains, such as fruits, seeds, wood etc., generally provide evidence of local presence, but are relatively sparse: their recognition has recently been much improved by techniques such as the scanning electron microscopy of seeds. By contrast, pollen is produced in vast amounts but its great mobility brings considerable difficultie n interpretation and in inferring local origin: in recent years, however, great advances have been made, not only in handling the statistical data but in applying high-power optical and electron microscopy to pollen recognition. These advantages are only now beginning to take serious effect, but by 1970 there was already a very substantial achievement to report (Godwin 1975).

The recurrent interglacial cycle of vegetation

By 1959 the conception had been adduced of a repeated cycle of interglacial change, in which, responding to climatic amelioration, the cold cryocratic conditions at the conclusion of a glacial stage, with unstable, immature, base-rich soils and open herbaceous vegetation with a strong arctic–alpine component, were displaced by warm protocratic conditions with stable soils supporting park-tundra or open woodland with steppe elements, 'weeds and ruderals' accompanying an immigrant thermophile flora. In turn followed a warmth maximum with brown-earth soils carrying closed deciduous forest with associated thermophilous herbs: this mesocratic phase was succeeded by a telocratic one characterized by leached soils and bog formation with spread of coniferous woodland and acidic heaths, prior to the refrigeration of the next glacial stage. This concept has been subsequently modified and refined, but no one doubts its general applicability to the interglacials of temperate western Europe, and it has been made the basis of the schematic division of British interglacials by Turner and West (1968) into a sequence of sub-stages: I Pre-temperate, II Early-temperate, III Late-temperate, IV Post-temperate. The use of these allows us to make comparison of vegetational events and floristic history between successive interglacial periods including our present Flandrian, with abundant data for the Ipswichian and Hoxnian and considerably sparser information for the Cromerian and still earlier interglacials. It is particularly evident in the pollen-analytic comparisons of protocratic, mediocratic, and

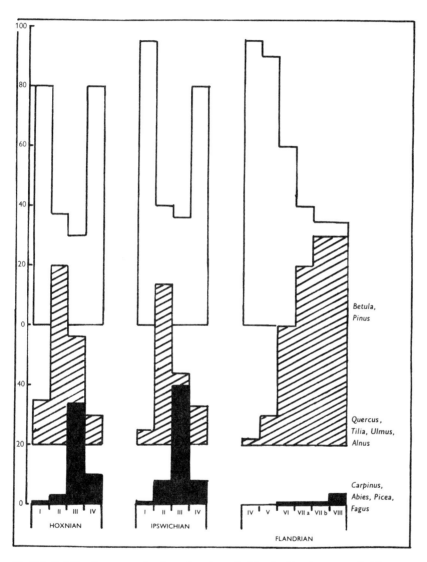

Fig. 9.1. Relative frequency of pollen of protocratic (*Betula, Pinus*), mesocratic (*Quercus, Tilia, Ulmus* and *Alnus*) and telocratic (*Carpinus, Abies, Picea, Fagus*) forest trees in the three latest interglacials: the inference appears to be that the Flandrian is an uncompleted interglacial stage.

telocratic tree genera in the three latest interglacials, as exhibited in Fig. 9.1, from which it seems evident that in the Flandrian we have so far only progressed to the early Late-temperate sub-stage.

The basis of the correspondence between these three interglacials is the overall control of the climatic progression, a control that gave the standard pollen-zones of the 'postglacial' their very important quasi-chronological function, a use from which they are being rapidly displaced by radiocarbon dating and to a lesser extent by other physical methods. However, within the pre-Flandrian interglacial stages to which radiocarbon dating cannot extend, the sub-stages I–IV, largely identified pollen-analytically, still have this major function, providing the coarse time-scale into which all sub-fossil records may be fitted, as well as giving direct evidence of the presence of taxa otherwise not identified.

During the glacial periods deposits yielding pollen are scarce and commonly contaminated with material derived from preceding interglacials, so that reconstruction of floristic events depends largely upon macroscopic plant remains that nevertheless can be locally abundant, well-preserved, and indicative of conditions conformable with the terminal interglacial stages and with the richer evidence of the interstadials that break up at least the latest glacial stage, the Devensian.

By the close of Tertiary time the bulk of plant species present in Britain were those that still live here or on the nearby continent. The Tropical, East Asian, and North American Tertiary elements had almost vanished and their place had been taken by the temperate European flora of today, with its heavy representation of herbaceous taxa. Successive Pleistocene stages give evidence of the final elimination of the Tertiary elements after longer or shorter persistence.

Early and Middle Pleistocene

Professor West has described our evidence for the plant life of early and middle Pleistocene, most notably at Ludham and through the 'Cromer Forest Bed Series'. As already shown by the analyses of Thomson at West Runton, the Cromerian interglacial forest history began with the dominance of pine and birch, showed a middle mesocratic phase dominated by *Ulmus*, *Quercus*, *Tilia*, and *Alnus* accompanied by consistently low frequencies of *Corylus*, and a telocratic phase characterized by very high values for *Picea* alongside increasing *Pinus*. At other Cromerian sites *Abies* accompanied these terminocratic elements. The warm sub-stage has produced such thermophilic arboreal genera as *Fraxinus*, *Viburnum*, *Frangula*, and *Juglans*, and, among the aquatics, *Najas minor*, *Salvinia natans*, *Azolla filiculoides*, and probably *Trapa natans*. These aquatics, like *Corema alba* and *Hypecoum procumbens*, also Cromerian, are no longer native to the British Isles.

The Hoxnian Interglacial

The second interglacial of the Middle Pleistocene is excellently defined at the type site at Hoxne and at Mark's Tey, Essex where annual laminations strongly indicate a total duration of *c*. 40 000 years. Other sites occur not only in East Anglia, but in the English Midlands, in the Shetland Islands, and most probably at Gort, Kilbeg, Baggotstown, and Kildromin in Ireland. In the East Anglian arboreal pollen diagrams sub-stage I is dominated by *Betula* and *Pinus*, and sub-stage II represents mixed-oak forest with *Alnus* and *Corylus* more frequent than in the Cromerian, but still not very abundant, with such thermophiles as *Hedera* and *Ilex*. The late-temperate sub-stage III shows early decrease of *Ulmus* and expansion of *Carpinus* and then *Abies*, perhaps foreshadowing the changes in sub-stage IV where there are high *Empetrum* values and dominance of *Pinus* and *Abies* with some *Betula*. The pollen both of *Buxus sempervirens* and of *Vitis* in sub-stage III indicates the high summer temperatures experienced. At Clacton, sub-stage III showed great expansion of *Abies*, but at Birmingham the dominant was *Picea*. The Irish Hoxnian sites, particularly Gort and Kilbeg, indicate a highly oceanic climate during late sub-stages III and IV where *Rhododendron ponticum* flourished with forests of largely evergreen character containing *Pinus, Abies, Picea, Taxus*, and *Buxus* beside the sparse mixed-oak forest elements. Here too, as in Fugla Ness (Shetland Islands) were many Ericaceae in the same part of the interglacial, including not only present-day Lusitanian components of the West Irish flora, such as *Daboecia cantabrica, Erica mackiana*, and *E. ciliaris*, but species such as *E. scoparia* no longer native. Besides those already mentioned, from England as well as Ireland, there are records of many plants no longer growing here, notably *Brasenia* cf. *purpurea, Lysimachis punctata, Nymphoides cordata*, as well as the Hiberno–American species *Eriocaulon septangulare*.

The Ipswichian Interglacial

Much more than for the Hoxnian, the sites for the Ipswichian Interglacial tend to be concentrated in East Anglia. The vegetational cycle in this interglacial differs in several respects from that of the Hoxnian, first in lacking the characteristic introductory phase of dominant *Hippophaë rhamnoides*. In sub-stage II *Alnus* is abundant much more locally, but *Acer* (probably *A. monspessulanum*) is a substantial forest component and *Corylus* accompanies the mixed-oak forest trees in high sustained frequencies. Again in sub-stage III, as the mesocratic deciduous forest elements decline *Carpinus* becomes quite important. The final sub-stage IV is again characterized by increased *Betula* and *Pinus* but with little sign of the *Picea* forests that go with them on the neighbouring European mainland. In contrast with the Hoxnian, the Ipswichian seems to exhibit vegetational response to a strongly continental climate as evidenced, for example, by the abundant fruiting of many aquatics such as *Lemna minor, Hydrocharis morsus-ranae*, and *Stratiotes aloides* together with *Najas minor, Trapa natans*, and *Salvinia natans*, which have since become extinct in Britain. The Irish Interglacial at Shortalstown contained a fruit of *Decodon*, a genus whose one living species is North American.

The Devensian glacial stage

Following the Ipswichian we have deposits of the one glacial stage, the Devensian (i.e., Weichselian), for which we have a substantial volume of plant records. Naturally enough the bulk of these come from south-eastern England, the region outside the limits of actual glaciation; they reflect periglacial conditions through the glacial stage and fortunately a few sites such as Wretton link the final Ipswichian with the opening Flandrian. Happily, Devensian deposits are mainly within the range of radiocarbon dating, and supply great numbers of macroscopic plant remains. These indicate that a generally treeless vegetation prevailed, very rich in open herbaceous and aquatic communities. There is strong representation of families, genera, and species that now have the status of ruderals or weeds, at least eighty recognizable at species level and inclusive of *Chenopodium, Atriplex, Plantago, Rumex, Ranunculus, Cirsium* spp. in great abundance, and such unexpected plants as *Centaurea cyanus, Pastinaca sativa*, and *Onobrychis viciifolia*. Plants of marshes and fens were equally well represented, notably among the carices, related genera such as *Scirpus, Eleocharis*, and *Eriophorum*. Junci were also abundant and there is increasing evidence of extensive open grasslands, habitat no doubt for the numerous large grazing mammalia. Aquatic plants too are richly represented, including many that, as with some of the terrestrial flora, are restricted in their northern range within Scandinavia. It is not surprising that the Devensian yields a long and striking list of plants of open montane and sub-arctic habitats, representative of the arctic–subarctic, arctic–alpine, and alpine categories. Most of these have withdrawn in the Flandrian to the mountains of the north and west, when indeed they have not been totally lost to the British

Isles: total loss has been extensive, particularly in Ireland, including *Betula nana, Astragalus alpinus, Minuartia stricta,* and *Koenigia islandica.* Of particular interest is the recurrent evidence for associations of halophytic plants, both obligate halophytes, such as *Glaux maritima, Juncus gerardii, Suaeda maritima,* and *Triglochin maritimum,* and facultative halophytes that include *Armeria maritima, Blysmus rufus, Eleocharis uniglumus, Najas marina,* and *Plantago maritima.* It seems probable that the necessary conditions were widespread permafrost and the great evaporation caused by a strongly continental climate, as in some sub-arctic regions of the present day. The presence of these halphytes also correlates with that of so-called 'steppe' species that include *Corispermum* sp. (no longer native), *Linum perenne, Helianthemum canum,* and *Artemisia* sp.: the actual presence of *Ephedra* sp. is still uncertain. In the Weichselian landscape, diversified by slope, drainage, aspect, snow-lie, etc., there was certainly scope for large communities of dwarf shrubs, particularly dominated by shrub willows or juniper, and on acid sites, by *Empetrum.* Locally leaching allowed the local development of acidicoles, with many characteristic Ericaceae such as *Calluna vulgaris, Erica tetralix,* and *Andromeda polifolia.*

The overall picture is one of considerable vegetational and floristic diversity outside the limits of the ice advances and it is remarkable that there is evidence, even now, for the presence of at least 20% of the present British Flora in Middle Devensian time, and as much as 30% in Late Devensian. More recording of fossil identifications will naturally increase these figures.

In view of the startling nature of these results it is interesting to consider the records for glacial stages preceding the Devensian, especially the opening and closing phases of the Wolstonian, Anglian, Beestonian, and Baventian. Despite their relative sparsity these records show remarkable similarity to those of the Devensian; again we encounter *Salix polaris, S. herbacea, Betula nana, Oxyria digyna, Saxifraga* spp., and the abundant pollen of genera such as *Centaurea, Plantago, Polygonum, Rumex, Scleranthus, Limonium,* and *Pastinaca.* It will be of great interest to see the extended lists for these periods.

Interstadial

The treelessness of the Devensian was broken during interstadia. The oldest of these, seen at Chelford in Cheshire, had a radiocarbon age of about 60 000 years and at its height exhibited woodland dominated by birch, pine, and spruce, the last appearance of *Picea* as a native British tree. Accompanying plant species and insect remains suggest equivalence with woodland now prevalent in Northern

Finland. In the West Midlands a complex of sites is aggregated in the Upton Warren Interstadial of Middle Devensian age: though milder than the general run of Devensian sites, these share the flora typical of the Devensian as a whole and show no evidence of carrying woodland. Finally in the Late Devensian the British Isles have yielded great numbers of sites exhibiting the triple character of the oscillation round the mild Allerød interstadial, dated here as on the European mainland between approximately 12 000 and 10 800 years BP, and set centrally between the 'postglacial' pollen-zones I and III. In zone II itself both pollen and macrofossils show the interstadial, over a large part of Britain, to have carried closed birch woodland with *B. pubescens* commoner than *B. pendula* and with some native *Pinus sylvestris* in southern and eastern England. A few woodland herbs accompanied this forest spread, and even in zone II its woodland character was modified in the north and west into communities dominated by grass, sedges, *Artemisia*, *Empetrum*, and *Juniperus*, according to local site and climate. There is some slight evidence for a wooded interstadial equivalent to the west European Bølling.

The Flandrian Interglacial

The major shifts in Flandrian forest history, established by pollen analysis, and following the pattern of interglacial climatic control are indicated in Fig. 9.2. We note that the earliest Flandrian was marked by widespread expansion of birch woodlands and of pine also in the south and east. It seems likely that even so early as this, *Corylus* was independently established on the mild west Scottish coast, not long preceding *Alnus glutinosa*. Many thermophilous plants now appeared and juniper scrub often marked the swift replacement of the glacial climate. There followed establishment and wide expansion of the elm and oak with hazel often vastly outstripping them in frequency, especially in the west, where hazel-scrub must have been prevalent. In England and Wales more particularly pine, woodland as well established, possibly with hazel undergrowth, and abundant thermophilous aquatics reflect the sharply improved climate and still eutrophic state of the lakes. The onset of the Atlantic period saw the total dominance, save for highland Scotland, of mixed oak forest, reinforced in England and Wales by the lindens, and generally by the alder, which expanded so suddenly over large areas that one has to suppose that a shift of climate had suddenly created a mosaic of wet glades and streamlets upon the forest floor. *Fraxinus* accompanied the shift and many stronger warmth indicators accompanied the hypsithermal, including *Hedera*, *Ilex* and *Viscum*, *Lonicera*, *Thelycrania*, and *Digitalis purpurea*.

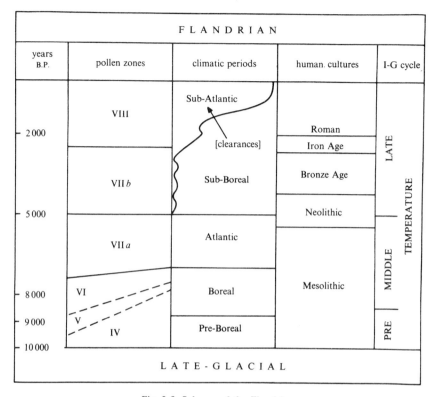

Fig. 9.2. Schema of the Flandrian.

Many of the thermophiles, especially the aquatics, can be shown by the fossil records now to have exceeded their present northern range. Pine forest was now established in Caledonia and has never subsequently been entirely displaced, and now began also the widespread growth of acidic raised-bog and blanket bog in western and upland areas.

Disforestation, survival, extinction, and re-invasion

About 5000 years BP, there began the processes of forest clearance and agriculture that distinguish Flandrian vegetational progression from that of earlier interglacials, and that complicate pollen-analytic and climatic interpretation even in respect of the telocratic forest genera, *Carpinus* and *Fagus*, that in England and Wales characterize the remainder of the Flandrian. From the earliest Neolithic onwards, scores of examples of forest clearance have been identified, very often with radiocarbon dating and conclusions as to the nature, purpose, and duration of the operation and of recovery from it.

The network of evidence extends into historic time, allows distinction of pasture and arable usage, of cultivation of crops such as *Secale*, *Cannabis*, and *Linum* and documents the great expansion of weeds, ruderals, and scrub, as well as indicating styles of woodland management and effects of soil erosion. Behind this pervasive pattern of anthropogenic effects we may, especially in peat-mires, see evidence for the Sub-atlantic climatic deterioration although here also drainage and peat-cutting obscure the picture.

The data-bank of Quaternary plant fossil records we have kept in Cambridge now has over 50 000 entries, a total allowing reconstruction in many particulars of the past history of the British flora (Godwin 1975). A substantial component, including many arctic–alpines but almost no trees, was present when Devensian ice-fields occupied the north. In the following Late Devensian there was very rapid

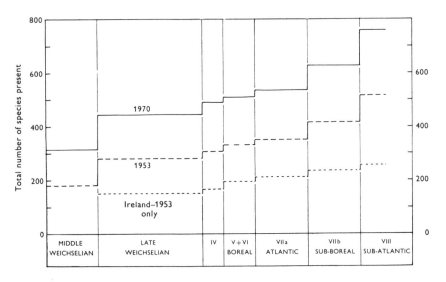

Fig. 9.3. Increment graph of all historically dated plant records for the British Isles from the middle Devensian. The comparable existing flora is about 1500 species. Broken lines show records for Ireland.

invasion across the North Sea (then dry) of a large part of our flora including many thermophiles, and this may have been paralleled by invasions from the continental shelf off our western shores. In the following early Flandrian the process continued so that by the time the North Sea had approached its present extent in the Atlantic period, the bulk of our flora, including our woodland plants, was established. However, some species later to establish themselves, such

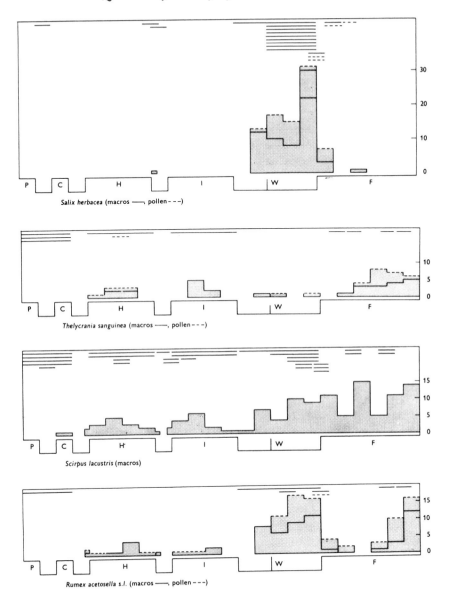

Fig. 9.4. Plant records, based on fossil pollen, or macros, or both, of four species of contrasting pattern, through all glacial and interglacial stages to the Late Flandrian. From above downwards (a) Glacial species (*Salix herbacea*), (b) exacting thermophiles (*Thelycrania sanguinea*), (c) fully persistent species (*Scirpus lacustris*), (d) 'Weeds' (*Rumex acetosella*). The Weichselian (Devensian) glacial stage is divided between Early- plus Middle-, and Late-.

as the lindens, beech, and hornbeam, never colonized Ireland. Subsequently the extension of clearances and plant husbandry not only provided for accidental introduction of weeds and ruderal plants, but allowed opportunity for those that had survived from Devensian times through the period of dense forest cover to re-expand prodigiously. Meanwhile the great extension of upland grazing and the spread of leaching and bog-growth at high altitudes, especially in the last 2000 years or so, had still further restricted the habitats of the groups of arctic–alpine species relict from the Devensian. It is not surprising that the present-day frequency of this element shows great decrease southwards, nor that we can trace the steepening of this gradient through the Flandrian records, as we can similarly show a gradient in the opposite direction for the continental and southern elements, no doubt favoured by extension of arable cultivation and clearance in the southern and eastern half of England.

The data-bank now has records for over half of the (macro) species of our present flora so that conclusions based upon it carry a great deal of weight, especially as many species have very full records. Naturally the Flandrian and Devensian supply the bulk of them, but none the less there is ample material to indicate that the processes recognized for the Flandrian (anthropogenic effects apart) operated similarly in each preceding interglacial. We find the northern elements preponderant in the early and late substages, with southern thermophilic elements present, often in large numbers in the middle of the interglacial. It is a matter of great interest how substantially the same species recur in the successive glacial or subglacial stages and there is the clearest indication that some have been present through several cycles up to the present day, never apparently encountering conditions that forced total extinction upon them. This of course is not to deny the evidence that during the enforced migrations and reinvasions of successive cycles, a number of species have been progressively lost from our flora, for example the tree genera *Pterocarya*, *Tsuga*, *Picea*, and *Abies*, *Acer monspessulanum* and *Vitis vinifera*, aquatics such as *Azolla filiculoides*, *Salvinia natans*, *Trapa natans*, *Brasenia purpurea*, *Nymphoides peltata* and ericoids such as *Rhododendron ponticum*, *Erica scoparia*, and *E. terminalis*. No doubt these changes have also much affected the gene structures of present taxa.

References

Godwin, H. (1975). *The History of the British Flora* (2nd. ed.). Cambridge University Press.

Turner, C. and West, R.G. (1968). The subdivision and zonation of interglacial periods. *Eiszeitalter Gegenw.* 19, 93.

10

The Flandrian forest history of Scotland: a preliminary synthesis

H. J. B. BIRKS

Abstract

The present forest patterns of Scotland are described. Oak forest with birch and hazel predominates in the south and west of Scotland. Pine forest with some birch and oak is characteristic of central and eastern Scotland. Birch forest is locally frequent in the north. Forest is absent from the far north and west including Orkney and Shetland, which experience the most severe climate. The available pollen analytical data from nineteen regions in Scotland for the last 10 000 years are summarized as generalized pollen diagrams. The forest history is reconstructed, and possible reasons for the observed regional differentiation are discussed in the light of the palynological data. Climatic, edaphic, anthropogenic, and historical factors have all influenced the past and present forest patterns of Scotland.

Introduction

In Scotland today, four major potential vegetation regions are deducable from existing woodland fragments (McVean and Ratcliffe 1962). South of the Grampian Highlands and up the west coast as far north as southern Skye oak forest with birch would predominate (see Fig. 10.1); in central Scotland pine forest would predominate with some birch and oak; to the north and west birch forest would predominate; and in exposed coastal areas in the far north, on the smaller Hebridean islands, and on Orkney and Shetland the landscape would be naturally treeless, even at low altitudes.

The existence of this pattern poses important ecological questions, answers to which may be provided by palaeoecological studies. When did the pattern originate and how did it develop? What factors controlled the vegetational differentiation? What were the relative importance of local environmental factors, plant migration, and man's activities in influencing the pattern? Has the pattern existed throughout Flandrian time, and have the boundaries between regions changed in position with' time?

This chapter reviews the available Flandrian pollen analytical evidence from Scotland and tentatively answers some of these questions. A series of generalized pollen diagrams from different areas are presented as a means of summarizing the major trends, both spatially and temporally, in the Flandrian pollen stratigraphy. However, it is necessary first to describe briefly the major forest types occurring in Scotland today, to provide a modern ecological background against which the palaeoecological evidence can be assessed.

Present forest patterns

Scottish forests have undergone considerable destruction over the last 5000 years. The extent is apparent from the maps prepared by McVean and Ratcliffe (1962) showing the actual and potential distributions of forest in Scotland. The potential distribution has been constructed from the distribution of existing natural and semi-natural woodlands, from surviving woodland fragments on ungrazed islands in lochs (McVean 1958), in ravines and other steep places, and from historical records, combined with ecological knowledge derived from Britain and from Scandinavia. This potential distribution is summarized in Fig. 10.1 which is based on Map 3 in McVean and Ratcliffe (1962).

South and west Scotland are potentially areas of oak forest with birch. The scarcity of relict pine and derived heather-moor supports the view that pine was unimportant in the natural forest. Surviving woodland fragments consist mainly of *Quercus petraea*† and *Betula pubescens* ssp. *odorata*. On basic soils *Ulmus glabra* and *Fraxinus excelsior* also occur, with *Corylus avellana* in more open areas. *Corylus* is not restricted to rich soils, and birch and hazel frequently form extensive stands where oak has been removed. Other trees and shrubs often found in Scottish oak woods include *Alnus glutinosa, Hedera helix, Ilex aquifolium, Lonicera periclymenum, Populus tremula, Prunus padus, P. spinosa, Rubus fruticosus, Salix cinerea, S. caprea, Sorbus aucuparia*, and *Viburnum opulus*. Many of the woods occur on steep boulder-strewn slopes and in the oceanic climate of the west, bryophytes, lichens, and filmy ferns cover the boulders and the trunks of the trees. Tittensor and Steele (1971) give an ecological account of the oakwoods of the Loch Lomond area, which is applicable to many oakwoods in southern Scotland.

On acid soils, the understorey of the oakwoods and the derived birch–hazel woods is generally species-poor. Under heavy grazing *Pteridium aquilinum* and *Deschampsia flexuosa* predominate, but with less grazing *Vaccinium myrtillus* is often abundant. Ungrazed woods have an understorey of *V. myrtillus, Luzula sylvatica*, and ferns such as *Dryopteris borreri, D. aemula*, and *Blechnum spicant*.

On more basic soils, the understorey of ungrazed woods is herb-rich, with little *Vaccinium myrtillus*. Typical species include *Anemone nemorosa, Anthoxanthum odoratum, Endymion non-scriptus, Holcus lanatus, Oxalis acetosella, Primula vulgaris*, and *Sanicula europea*. With heavy grazing, this understorey becomes one dominated by grasses and *Pteridium aquilinum*. With complete felling, accompanied by grazing, species-rich *Agrostis-Festuca* grassland is produced.

The extension of oakwoods up the west coast of Scotland (Fig. 10.1) has some climatic significance, as many climatic factors involving warmth have a northeast–southwest gradient which appears to become critical near the Isle of Skye (Climatological Atlas 1952). This is reflected in plant distribution patterns as many southern Atlantic species with Macaronesian-Tropical affinities (Greig-Smith 1950) reach their northernmost world localities in this area; for example, the liverworts *Adelanthus decipiens* (see Fig. 8 in Ratcliffe 1968), *Jubula hutchinsiae*, and *Radula voluta*, and the filmy fern *Hymenophyllum tunbrigense*.

The Central Highlands and the east coast of Scotland are potentially areas of *Pinus sylvestris*-dominated forest, often with birch (mainly *B. verrucosa*), and with *Quercus petraea* in sheltered situations.

†Plant nomenclature follows Clapham, Tutin, and Warburg (1962) for vascular plants and Paton (1965) for hepatics.

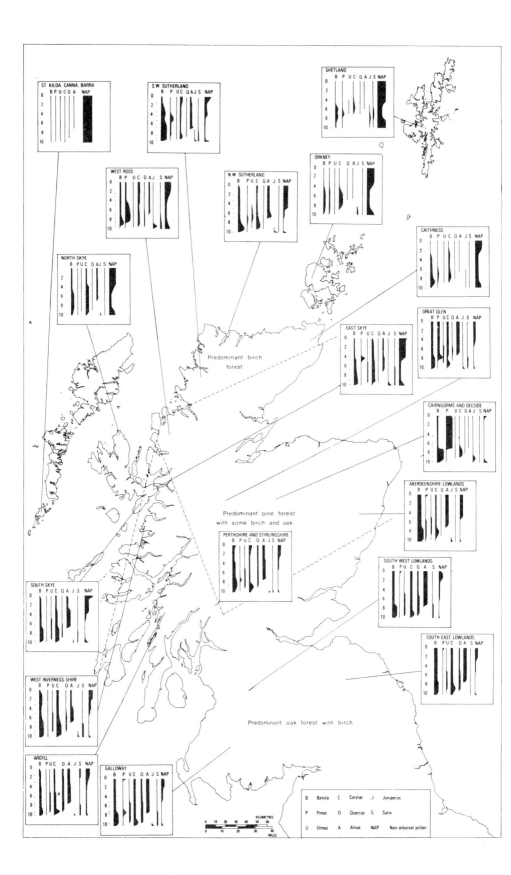

ST KILDA, CANNA, BARRA
B P U C Q A   NAP

S.W. SUTHERLAND
B P U C   Q A J S NAP

SHETLAND
B P U C Q A J S NAP

WEST ROSS
B P U C Q A J S NAP

N W SUTHERLAND
B   P U C Q A J S NAP

ORKNEY
B P U C   Q A J S NAP

CAITHNESS
B P U C Q A J S NAP

NORTH SKYE
B P U C   Q A J S NAP

EAST SKYE
B P U C   Q A J S NAP

GREAT GLEN
B P U C Q A J S NAP

Predominant birch forest

CAIRNGORMS AND DEESIDE
B P U C Q A J S NAP

ABERDEENSHIRE LOWLANDS
B P U C Q A J S NAP

Predominant pine forest
with some birch and oak

PERTHSHIRE AND STIRLINGSHIRE
B P U C Q A J S NAP

SOUTH WEST LOWLANDS
B P U C Q A S NAP

SOUTH SKYE
B P U C Q A J S NAP

SOUTH EAST LOWLANDS
B P U C Q A S NAP

WEST INVERNESS-SHIRE
B P U C   Q A J S NAP

ARGYLL
B P U C Q A J S NAP

GALLOWAY
B P U C Q A J S NAP

Predominant oak forest with birch

KILOMETRES
0 10 20 30 40 50 60
0   10   20   30   40
MILES

B Betula	C Corylus	J Juniperus
P Pinus	Q Quercus	S Salix
U Ulmus	A Alnus	NAP Non-arboreal pollen

Steven and Carlisle (1959) have mapped and described the known native Scottish pinewoods.

In the eastern pinewoods, the understorey varies with the density of the tree canopy. In dense pinewood, tall shrubs are rare, although *Betula verrucosa* and *Sorbus aucuparia* may occur in the canopy. The ground flora is dominated by *Vaccinium myrtillus* and *V. vitis-idaea*, with some *Calluna vulgaris*. In more open forest, *Juniperus communis* is characteristically abundant, and the ground flora is dominated by tall *C. vulgaris* and *V. myrtillus* associated with abundant hypnaceous mosses. Many of the pinewoods have been disturbed by felling and replanting.

The western pinewoods also contain abundant tall *J. communis* and *C. vulgaris*, with thick carpets of hypnaceous mosses and *Sphagnum* spp. *Ilex aquifolium*, *Sorbus aucuparia*, and *Prunus padus* may occur with *Betula pubescens* and *B. verrucosa* in the canopy.

Birchwoods (*B. verrucosa* with some *B. pubescens*) are widespread in central Scotland, often occupying fertile soils but also recolonizing openings in the acid pinewoods. Here their understorey resembles that of pinewoods, but on more fertile soils the understorey is herb-rich, with *Potentilla erecta, Oxalis acetosella, Melampyrum pratense,*

Fig. 10.1. Summary pollen diagrams from different areas of Scotland in relation to the potential forest zones proposed by McVean and Ratcliffe (1962). An approximate time-scale in thousands of years BP is shown for each diagram. For further details see text. The published diagrams used for each area are as follows:

Shetland: Hawksworth (1969), Johansen (1975).
Orkney: Moar (1969a).
Caithness: S. Peglar (unpublished).
North-west Sutherland: H.H. Birks (unpublished).
South-west Sutherland: Moar (1969b), H.H. Birks (unpublished).
West Ross: Birks (1972b), Durno and McVean (1959), Pennington et al. (1972).
West Inverness-shire: W. Williams (unpublished).
Southern Argyll: Donner (1957), Nichols (1967), S. Peglar (unpublished), Rymer (1974).
South-west lowlands: Durno (1965), Fraser and Godwin (1955), Nichols (1967).
Galloway: Birks (1972a, 1975), Moar (1969c), Nichols (1967).
Cairngorms and Deeside: Birks (1970, 1975), B. Huntley (unpublished), R.W. Mathewes and H.H. Birks (unpublished), O'Sullivan (1973, 1974a), Pears (1968, 1972), Vasari and Vasari (1968).
Aberdeenshire lowlands: Durno (1956, 1959, 1961, 1965, 1970), Fraser and Godwin (1955), B. Huntley (unpublished), Knox (1954), Vasari and Vasari (1968).
South-east lowlands: Newey (1968).
Great Glen: Pennington et al. (1972).
Perthshire and Stirlingshire: Brooks (1972), Donner (1957, 1962), Durno (1956), Newey (1966), Vasari and Vasari (1968).
Northern Skye: Birks (1973), Vasari and Vasari (1968), W. Williams (unpublished).
Southern Skye: Birks (1973), W. Williams (unpublished).
Eastern Skye: W. Williams (unpublished).
St. Kilda, Canna, Barra: Blackburn (1946), Flenley and Pearson (1967), McVean (1961).

and *Anthoxanthum odoratum*. Clearance of pine or birch forest followed by heavy grazing and burning leads to *Calluna* moor or species-poor *Agrostis-Festuca* grassland.

Oak occurs in scattered localities within the Central Highlands, generally on low ground and often on south-facing slopes. Its distribution tends to complement pine, suggesting that the balance is influenced by geology, topography, climate, and altitude.

The altitude and composition of the natural tree-line in Scotland are unknown, due to the virtual elimination of sub-alpine scrub communities by clearance, burning, and grazing. Three main types can be recognized from local remnants. Sub-alpine juniper scrub rich in ferns occurs most frequently above 610 m in the Eastern Highlands (McVean and Ratcliffe 1962). Natural transitions above pinewoods can be seen locally. *Betula nana* is absent from this community, in contrast to its behaviour in Norway. Sub-alpine birchwoods (*B. pubescens* ssp. *odorata*) are rare in the Central Highlands between 500 and 610 m and often have an abundance of juniper and tall herbs in the understorey. Sub-alpine willow scrub occurs on mildly basic soils between 670 and 910 m, and is dominated by *Salix lapponum, S. lanata, S. arbuscula*, or *S. myrsinites*. The fragmentary stands are usually on inaccessible cliff ledges protected from grazing.

There is evidence for the altitudinal descent northwards of the potential tree-line (Spence 1960) from about 750 m in the Central Highlands, 400 m in the north-west Highlands, to about sea-level in the Shetlands, associated with the progressively more severe climate at equivalent altitudes. There is a downward shift in the altitude of all the vegetation zones, including cultivated land.

Northern Scotland and the Western Isles are potentially areas of birch forest. Exposure prevents the growth of all trees except birch and hazel, and alder and willows in wet habitats. Herb-rich hazel scrub, often severely wind-pruned, frequently occurs on the talus below steep sea-cliffs that are inaccessible to grazing animals. Birch woods (mainly *B. pubescens* spp. *odorata*) grow on the more acid soils, and often contain *Ilex aquifolium, Sorbus aucuparia, Prunus padus, Salix cinerea*, and *Lonicera periclymenum*. Their understorey resembles that of oak and birchwoods in south and west Scotland (see Birks 1973). Bryophytes and lichens luxuriate on boulders and on the tree trunks and lower branches.

Other forest trees lack such a clear regional pattern within Scotland. *Fraxinus excelsior* is limited to good soil, and may form the natural woodland on shallow limestone soils south of Sutherland (McVean and Ratcliffe 1962). *Alnus glutinosa* is widespread throughout Scotland in suitable habitats. McVean (1956) describes some

Scottish alderwoods, and shows that the understorey depends upon the type of vegetation present before alder colonization. Alder can accompany willow and birch in swampwoods and eventually replaces them owing to its greater longevity and competitive power. It also forms woods on flushed hillsides and in hollows within woods, generally on richer soils. The understorey varies with the water and base content of the soil.

On the exposed coasts of Sutherland and Caithness, on Orkney and Shetland, and on many of the Hebridean islands, forest never developed extensively. *Salix aurita*, *S. cinerea*, and *S. caprea* scrub may have been important in some areas, with *Rubus fruticosus*, *Rosa* spp., and *Lonicera periclymenum* (Spence 1960), whereas maritime grasslands and heaths may have predominated in the most exposed areas.

Flandrian forest patterns

Summary pollen diagrams for nineteen areas within Scotland are presented in Fig. 10.1 as a means of synthesizing what is known at the time of writing (December 1975) about Scottish Flandrian forest history. Curves for the major tree and shrub pollen types and for total non-arboreal pollen are shown. The diagrams are based on percentages of total pollen excluding obligate aquatics, and are drawn on a time scale from 0 to 10 000 years ago, based on the radiocarbon dates shown as dots beside the relevant diagram. Where only a few dates are available for an area, linear interpolation or extrapolation provides an approximate time-scale. Some areas lack radiocarbon dates, and the tentative time-scale shown is derived by comparison of the pollen stratigraphy with nearby areas. The summary diagrams are based, as far as possible, on pollen sequences from lochs rather than from bogs to try to avoid local influences. If necessary, obviously local influences have been eliminated.

In the 'predominant oak forest with birch' area, six summary pollen diagrams are presented (Fig. 10.1). After a phase of *Juniperus communis* and herb dominance around 10 000 BP, *Betula* (*B. pubescens* and *B. verrucosa*) and *Corylus avellana* expanded at about 9 700 BP to form birch-hazel woodland, containing *Populus tremula*, *Sorbus aucuparia*, *Salix* spp., and *Prunus padus*. The date of the *Corylus* expansion has no obvious pattern. Birch-hazel woodland was invaded by oak (? *Quercus petraea*) and elm (probably *Ulmus glabra*) at about 8 500 BP to form mixed deciduous forest, with *Hedera helix*, *Ilex aquifolium*, *Sorbus aucuparia*, *Viburnum opulus*, and *Lonicera periclymenum*. Ferns were common, and spores of the filmy fern *Hymenophyllum wilsonii* have been recorded, emphasizing the oceanic climate. *Pinus sylvestris* arrived at about 8000 BP, but remained rare, occupying marginal habitats, such as dried peat-bog surfaces in Galloway between about 7500 and 6800 BP where the pine stumps were subsequently buried by peat growth (Birks 1975). *Tilia cordata*, to judge by its fossil pollen values, was absent from the area, and *Fraxinus excelsior* was rare, perhaps restricted to damp, mildly basic openings in the forest. *Alnus glutinosa* largely replaced *Salix* in wet habitats. The chronology and dynamics of its expansion are not yet clear, but its main increase in north-west England occurred at about 7500 BP (Smith and Pilcher 1973), compared with about 7000 BP in Galloway and the southern Scottish lowlands, 6800 BP in Argyll, and about 6500 BP further north in western Inverness-shire and southern Skye. However, there is convincing evidence for local growth of alder in southern Argyll from at least 7985 BP prior to its regional expansion there (Rymer 1974).

The decline in *Ulmus* pollen at 5000 BP marked the beginning of human interference in the area. The forests suffered reduction, particularly in the last 2000 years, until the largely treeless condition of the present day was reached. During recent time conifers have been widely planted, and an increase in conifer pollen is detectable in pollen diagrams that extend to the present day.

All the pollen diagrams from this area indicate that oak forest with birch was the natural forest vegetation of the area, at least as far north as Argyll and Perthshire. The altitude of the tree-line is not known, but in the Galloway hills *Betula*, *Populus tremula*, and *Sorbus aucuparia* may have occurred up to at least 610 m altitude (Birks 1972a). Although important before 5000 BP, *Ulmus* was reduced by man's activities and by progressive soil deterioration, leaving *Quercus*, *Corylus*, and *Betula*, with *Alnus* and *Salix* on wetter sites as the major forest components.

Summary pollen diagrams are available from four areas within the region of 'predominant pine forest with some birch and oak'. The vegetational history of the important native pinewoods of Deeside and the Cairngorms is illustrated by several recent pollen diagrams. After a phase of herb dominance, *Betula* (mainly *B. verrucosa*) and *Juniperus communis* expanded at 9800 BP accompanied by *Populus tremula* and *Sorbus aucuparia* to form an open-canopied woodland. At 8600 BP *Corylus avellana* expanded but it never achieved values as high as it did at this time in the south or west of Scotland. *Quercus* and *Ulmus* were always rare, and *Fraxinus* was probably absent, presumably because of the acidic soils and the cool continental climate of the area. At about 7000 BP *Pinus sylvestris* arrived and expanded rapidly to form the pine forest which still survives in parts today. *Alnus glutinosa* pollen is present from about 7000 BP and expands slightly between about 6500 and 6000 BP. Its values are so low, however, that it was clearly a rare tree in the Cairngorms, perhaps restricted to wet, minerotrophic situations. Evidence for human activity in the lowlands is slight, with some forest clearance dating from 3600 BP (O'Sullivan 1974*a*, *b*). More extensive forest clearance occurred in the hills, leading to the widespread development of heather moor in the last 2000 years (Birks 1975).

The pollen evidence shows the dominance of *Pinus sylvestris* on the acid well-drained soils of the Cairngorm area, and the rarity of other trees such as *Ulmus*, *Quercus*, and *Alnus* throughout the Flandrian. *Betula*, *Populus tremula*, *Sorbus aucuparia*, and *Prunus padus* may have been locally frequent in some situations. The altitudinal limit of forest is not known in detail but pine and birch remains have been found in blanket peats up to 790 m altitude (Pears 1968, 1972). Judging by modern tree-lines in Scandinavia, it is likely that pine and birch may have formed the natural upper limit in the Cairngorms, depending on local conditions, and were probably fringed by a belt of *Juniperus* scrub below dwarf-shrub heath and summit vegetation.

To the east of the Cairngorm area, pollen diagrams from the Aberdeenshire, Nairnshire, Kincardineshire, and Angus lowlands suggest that *Betula* dominated over *Pinus sylvestris* and *Corylus avellana*, and, after 6500 BP, *Alnus glutinosa*, *Ulmus*, and *Quercus* were rare. The success of *Betula* over *Pinus* may be related to the more fertile soils in the lowlands. Interestingly McVean and Ratcliffe (1962 Map B) suggest that much of the forest of the coastal lowlands may have been predominantly *Betula* rather than *Pinus*. Forest clearance was extensive from about 5000 BP but detailed palynological investigations in relation to the local archaeological history are required before man's influence on the woodlands can be fully elucidated.

To the west of the Cairngorm area, there is a single pollen diagram from the Great Glen (Pennington *et al.* 1972). This area is regarded by McVean (1964) to have supported dominant oak forest to about 300 m, with pine and birch above. The pollen diagram indicates that after a phase of *Betula* and *Juniperus communis* dominance, *Corylus avellana*, *Ulmus*, and *Quercus* expanded to form a mixed deciduous woodland. Between about 7800 and 6500 BP *Pinus sylvestris* increased, presumably forming pine forest on poorer soils and at higher altitudes than the deciduous forest. *Alnus* expanded slightly later and *Pinus* decreased somewhat. The combination of moderately high *Pinus*, *Betula*, *Corylus*, *Alnus*, and *Quercus* pollen percentages up to the present day indicates the complex forest composition of the

Great Glen area. Additional studies are required before its forest history can be confidently reconstructed.

Further west in West Ross where magnificent pine woods survive today in the Loch Maree area, several pollen diagrams are available (Birks 1972*b*; Pennington *et al.* 1972). After about 9800 BP *Juniperus communis* formed widespread fern-rich juniper scrub. *Betula* and *Corylus avellana* expanded at about 8900 BP to form open birch-hazel woods with *Sorbus aucuparia*, ferns, and tall herbs such as *Trollius europaeus*, *Cirsium*, and *Filipendula*.

At about 8250 BP in southern West Ross, and about 7900 BP in northern West Ross, *Pinus sylvestris* replaced birch on all soils except the wettest and the most fertile. Ferns were more important in these western pine forests than in the Cairngorms, but *Ulmus* and *Quercus* were just as rare here as they were in the Cairngorms and west Inverness-shire. However, *Alnus glutinosa* became common at about 6500 BP. *Pinus* pollen started a slow decline at about 7000 BP, perhaps as a result of soil deterioration in the oceanic climate of the area (Birks 1972*b*). However, it decreased sharply at about 4000 BP. The reasons for this widespread spectacular decline in pine pollen at this time in northwest Scotland are not clear but a combination of climatic changes and human activity including burning may have initiated the replacement of pine forest on flat and gently sloping ground by treeless blanket-bog. There is independent evidence from chemical analyses of lake sediments (Pennington *et al.* 1972) and from peat stratigraphy (Lamb 1964; Birks 1975) to indicate a change to a more oceanic climate with strong winds and increased precipitation and cloudiness at this time. This would inhibit the regeneration of pine on mineral soils by reducing the number of good seed years and would also cause waterlogging to be more widespread and hence encourage the expansion of blanket-bog. The demise of pine at about 4000 BP is not restricted to Scotland, for Smith and Pilcher (1973) describe a comparable pine decline in north-east Ireland between 4400 and 3800 BP. The expansion of blanket-bog at the expense of forest continued to the present day, thereby forming the characteristic blanket-bog landscape of West Ross. Pine and birch woods survive only in rocky areas too steep for bog to develop.

The four summary diagrams within the 'predominant pine' zone show variation, both in the relative proportions of different tree types, particularly birch and pine, and in the timing of the expansion of pine in the west (8300 BP) and in the east (7000 BP). More dated pollen sequences are required from the Central Highlands for the forest patterns to be more fully understood, but present evidence suggests that the oldest pine forests in Scotland are in West Ross rather than in the Cairngorm area.

Pollen diagrams are available from three areas within the northern 'predominant birch forest' zone. Pollen analyses from south-west Sutherland indicate the replacement of *Juniperus communis*-dominated scrub by *Betula* at 9200 BP and the subsequent expansion of *Corylus avellana* at 8950 BP to form fern- and herb-rich birch-hazel woods containing *Populus tremula, Prunus padus, Sorbus aucuparia,* and *Viburnum opulus*. While *Corylus* expanded in southern Sutherland, *Juniperus*-dominated vegetation was being replaced by *Betula* woodland in parts of West Ross, illustrating the complex chronology of the early Flandrian forest succession in Scotland. In Sutherland a small rise in *Pinus sylvestris* pollen values at 8300 BP probably reflects long-distance dispersal of pine pollen from the south. At 6500 BP *Alnus glutinosa* expanded and at 5300 BP *Pinus* expanded locally. This phase of pine growth ceased at 4000 BP, as it did in West Ross, probably due to the spread of blanket-bog and associated deforestation. More extensive deforestation of birchwoods occurred at about 1500 BP. Scattered birchwoods survive today on steep rocky slopes and in ravines in the otherwise treeless Sutherland landscape. There is evidence that *Quercus* was never important, although *Ulmus* may have occurred locally prior to 5000 BP, perhaps on and near the outcrops of Durness limestone. Other thermophilous trees and shrubs such as *Fraxinus excelsior, Ilex aquifolium*, and *Lonicera periclymenum* may also have occurred locally.

In north-west Sutherland, the pollen stratigraphy lacks the expansion of *Pinus* pollen at about 5300 BP. After a brief phase of *Juniperus* dominance, *Betula* and, subsequently, *Corylus avellana* expanded to form herb and fern-rich birch-hazel woods similar to those in south-west Sutherland. *Ulmus* was present on the Durness limestone. At about 6000 BP *Alnus* became abundant locally but neither *Pinus* nor *Quercus* were ever important. *Hedera helix* and *Ilex aquifolium* were both present locally. Deforestation occurred progressively from about 5000 BP, and there is palynological evidence to indicate both pastoral and arable farming from this time to the present day.

To the east in Caithness, the available pollen evidence suggests that after a phase of herbaceous dominance, *Betula* expanded at about 9300 BP along with *Corylus* to form local patches of birch-hazel woodland, presumably in sheltered situations. Tall-herb and fern-dominated communities with *Salix* may have been widespread. The low values of *Pinus, Ulmus, Quercus*, and *Alnus* pollen indicate that no extensive closed forest ever developed in Caithness, although oak, elm, and alder grow there today in sheltered ravines, and fossil pine stumps are frequent in the extensive blanket-bog country in western Caithness and eastern Sutherland. Clearance of what woodland there was in Caithness occurred from 3000 BP and as in north-west Sutherland there is evidence for both arable and pastoral farming.

The three summary diagrams from the 'predominant birch' zone all indicate that birch and hazel have always been the dominant trees although there is considerable variation in the density of forest, with scattered woodland in Caithness and more extensive forest in Sutherland. The importance of pine varies too. It once grew north of its present presumed native limit between 5300 and 4000 years ago in south-west Sutherland, but never as far north as north-west Sutherland. This short *Pinus* phase contrasts with the long period of pine dominance (4000–8300 BP) in West Ross.

Proceeding northwards to Orkney and Shetland, the available pollen evidence suggests that no extensive woodland ever developed during the Flandrian. On Orkney Mainland, *Betula* and *Corylus* scrub probably developed locally, as in Caithness. The predominant vegetation in mid-Flandrian times was tall-herb communities with *Rumex acetosa, Filipendula, Urtica,* and ferns. It is unlikely that *Ulmus, Quercus,* or *Alnus* ever occurred on Orkney. The status of *Pinus* is more problematical (see Moar 1969*a*). In Shetland pioneer herbaceous communities with *Rumex acetosa, R. acetosella, Sedum,* and *Oxyria digyna* persisted to about 9000 BP, after which some *Betula* (including *B. nana* and *B. pubescens*) and *Corylus avellana* scrub developed locally along with *Salix* and *Juniperus communis.* Tall-herb vegetation was the most widespread type with *Filipendula* and ferns. From about 5000 BP there was a reduction in the amount of woodland, possibly due to man's activities, to grazing animals, and to soil deterioration and the spread of blanket-bog. There is no palaeobotanical evidence for the growth of *Alnus, Quercus, Ulmus,* or *Pinus* on the Shetlands.

There are, at present, sparse pollen analytical data from the Hebrides, with the exception of the Isle of Skye. Pollen diagrams from small Hebridean islands (Canna, Barra, and St. Kilda) all indicate that no tree development occurred during the Flandrian, presumably because of intense exposure to westerly gales. The main vegetation was maritime grasslands and dwarf-shrub heaths with *Plantago maritima, P. lanceolata,* and *Armeria maritima.*

Pollen diagrams covering the entire Flandrian are available from three areas of contrasting geology and present vegetation within the Isle of Skye, namely southern Skye, eastern Skye, and northern Skye (regions 1, 2 and 6 respectively on Fig. 5 in Birks 1973). In southern

Skye the pollen stratigraphy is similar to that from the adjacent mainland and suggests that *Betula* (mainly *B. pubescens*) and *Corylus avellana* were the predominant forest trees from about 9600 to 6500 BP although *Ulmus, Populus tremula, Sorbus aucuparia, Salix*, and *Viburnum opulus* also occurred locally. *Alnus glutinosa* became an important local component at about 6500 BP. The *Quercus* pollen values are lower on Skye than on the adjacent mainland, suggesting that the northern limit of oak was near here. This apparent rarity of oak on Skye contradicts the hypothesis, based on floristic and ecological grounds (Birks 1973), that southern Skye formerly supported oak woodland. As in the south, *Pinus* was unimportant in the forests of southern Skye, contrasting with nearby West Ross and the Torridonian sandstone area of eastern Skye. *Corylus avellana* expanded early in Skye (about 9600 BP) compared with its expansion in West Ross and Sutherland.

The pollen stratigraphy from eastern Skye links the areas to the south and to the north. The early Flandrian *Juniperus communis* and herb phase was replaced by *Betula* and *Corylus avellana* with *Sorbus aucuparia, Populus tremula*, and *Prunus padus* as in southern Skye. *Alnus glutinosa* expanded at about 6500 BP. In contrast to sites in the south, *Pinus* pollen had a short period of abundance between about 4600 and 4000 BP. This may reflect the local growth of pine on dried peat surfaces, a widespread phenomenon in north-west Scotland at this time (Birks 1975). The sharp decline in *Pinus* pollen corresponds to the widespread pine decline in north-west Scotland at about 4000 BP. *Betula, Corylus*, and *Alnus* also decline at this time, probably reduced by the spread of blanket-bog in this part of Skye.

In the wind-exposed basaltic area of northern Skye, the available pollen data suggest that closed woodland never developed during the Flandrian. From about 9500 to 6500 BP *Betula, Corylus avellana, Populus tremula*, and *Sorbus aucuparia* occurred locally, presumably in sheltered situations. Species-rich grassland and tall-herb vegetation were frequent in the moist, fertile basalt lowlands. *Alnus glutinosa* expanded at about 6500 BP but was never abundant. From about 5000 BP deforestation proceeded, resulting in the characteristically treeless landscape of northern Skye today.

The Flandrian forest history of Skye corresponds closely with the present distribution of woodland fragments on the island (see Birks 1973). Southern Skye has the most woodland today. In eastern Skye, woodland is restricted to rocky slopes too steep for blanket-bog development, and in northern Skye trees only grow in sheltered places.

This pattern may reflect climatic differences within Skye, southern

Skye being the mildest part today (Birks 1973). There is considerable geographical variation in the Flandrian forest history of Skye, just as there is in the Late-Devensian vegetational history and in the present flora of the island (Birks 1973). The forest history of southern Skye has the greatest affinities with sites in the 'predominant oak forest with birch' zone of southern and western Scotland. The history of northern Skye corresponds closely with sites in the 'predominant birch forest' areas of the north, and the history of eastern Skye has some affinities with parts of West Ross in the 'predominant pine forest with some birch and oak' region. Nowhere else in Scotland can this range of variation in forest history be found within such a small area.

Conclusions

The available pollen data suggest that the major forest patterns in Scotland were established by about 6000 BP. The causes for the differentiation seem to have been a complex interaction between climate and soils coupled with different times of immigration of the various tree taxa. In the relatively mild south and west, as far north as southern Skye, *Corylus* and *Betula* had become established by about 9000 BP. On its arrival *Pinus* could not compete successfully with *Corylus* or later with *Quercus* and *Ulmus* which eventually replaced *Corylus*. Only in the Galloway hills was *Pinus* able to gain a local foothold by occupying dried peat surfaces between 6800 and 7500 BP.

In the Cairngorms *Corylus*, *Quercus*, and *Ulmus* were unable to flourish on the poor acid soils and in the relatively cold continental climate, unlike *Betula*. On its arrival, *Pinus* largely replaced *Betula* to form the pine forests that persist today. Further east in the Aberdeenshire lowlands, more fertile soils allowed *Betula* and *Corylus* to grow well and thus to exclude *Pinus* from much of the area. *Ulmus* and *Quercus* were rare because of the unfavourable climate.

In West Ross, north of the limit of *Quercus*, *Pinus* forest developed extensively on both mineral soils and peats overlying the acid rocks. The extremely wet climate coupled with man's activities resulted in the expansion of blanket-bog at the expense of the pine forests. To the north *Betula* and *Corylus* were the only trees sufficiently well adapted to the gale frequencies to form areas of woodland, although there is some evidence to indicate that *Pinus* expanded both westwards to Skye and northwards to south-west Sutherland about 5000 years ago.

Man has had considerable effects in reducing the extent of woodland in Scotland over the last 5000 years. The few surviving fragments of natural and semi-natural woodland merit conservation not only as ecosystems of scientific interest but as living representatives of the former forests of Scotland.

Acknowledgements

Much of the work reviewed here has been done during the tenure of a research grant from the Natural Environment Research Council. This support is gratefully acknowledged. I am greatly indebted to my friends and colleagues Mrs Sylvia Peglar, Dr Rolf Mathewes, Dr Leslie Rymer, Dr Brian Huntley, and Dr Willie Williams and to my wife Dr Hilary H. Birks for generously allowing me to incorporate their unpublished results in this review. Discussions with these persons and with Dr W. Tutin and Dr D.A. Ratcliffe have helped clarify several of the ideas presented here.

The manuscript has benefited from critical reading by Dr Hilary H. Birks.

References

Birks, H.H. (1970). Studies in the vegetational history of Scotland I. A pollen diagram from Abernethy Forest, Inverness-shire. *J. Ecol.* **58**, 827-46.

Birks, H.H. (1972*a*). Studies in the vegetational history of Scotland II. Two pollen diagrams from the Galloway Hills, Kirkcudbrightshire. *J. Ecol.* **60**, 183-217.

Birks, H.H. (1972*b*). Studies in the vegetational history of Scotland III. A radio-carbon-dated pollen diagram from Loch Maree, Ross and Cromarty. *New Phytol.* **71**, 731-54.

Birks, H.H. (1975). Studies in the vegetational history of Scotland IV. Pine stumps in Scottish blanket peats. *Phil. Trans. R. Soc.* **B 270**, 181-226.

Birks, H.J.B. (1973). *Past and present vegetation of the Isle of Skye—A palaeoecological study*. Cambridge University Press, London.

Blackburn, K.B. (1946). On a peat from the island of Barra, Outer Hebrides. Data for the study of post-glacial history X. *New Phytol.* **45**, 44-9.

Brooks, C.L. (1972). Pollen analysis and the Main Buried Beach in the western part of the Forth Valley. *Trans. Inst. Br. Geogr.* **55**, 161-70.

Clapham, A.R., Tutin, T.G. and Warburg, E.F. (1962). *Flora of the British Isles*. Cambridge University Press, London.

Climatological Atlas of the British Isles (1952). H.M.S.O., London.

Donner, J.J. (1957). The geology and vegetation of late-glacial retreat stages in Scotland. *Trans. R. Soc. Edinb.* **63**, 221-64.

Donner, J.J. (1962). On the post-glacial history of the Grampian Highlands of Scotland. *Soc. Scient. Fennica Comm. Biol.* **24**, 29pp.

Durno, S.E. (1956). Pollen analysis of peat deposits in Scotland. *Scott. geogr. Mag.* **72**, 177-87.

Durno, S.E. (1959). Pollen analysis of peat deposits in the eastern Grampians. *Scott. geogr. Mag.* **75**, 102-11.

Durno, S.E. (1961). Evidence regarding the rate of peat growth. *J. Ecol.* **49**, 349-51.

Durno, S.E. (1965). Pollen analytical evidence of 'landnam' from two Scottish sites. *Trans. Proc. bot. Soc. Edinb.* **40**, 13-19.

Durno, S.E. (1970). Pollen diagrams from three buried peats in the Aberdeen area. *Trans. Proc. bot. Soc. Edinb.* **41**, 43-50.

Durno, S.E. and McVean, D.N. (1959). Forest history of the Beinn Eighe Nature Reserve. *New Phytol.* 58, 228–36.

Flenley, J.G. and Pearson, M.C. (1967). Pollen analysis of a peat from the island of Canna (Inner Hebrides). *New Phytol.* 66, 299–306.

Fraser, G.K. and Godwin, H. (1955). Two Scottish pollen diagrams: Carnwath Moss, Lanarkshire, and Strichen Moss, Aberdeenshire. Data for the study of post-glacial history. XVII. *New Phytol.* 54, 216–21.

Greig-Smith, P. (1950). Evidence from hepatics on the history of the British flora. *J. Ecol.* 38, 320–44.

Hawksworth, D.L. (1969). Studies on the peat deposits of the island of Foula, Shetland. *Trans. Proc. bot. Soc. Edinb.* 40, 576–91.

Johansen, J. (1975). Pollen diagrams from the Shetland and Faroe islands. *New Phytol.* 75, 369–87.

Knox, E.M. (1954). Pollen analysis of a peat at Kingsteps Quarry, Nairn. *Trans. Proc. bot. Soc. Edinb.* 36, 224–29.

Lamb, H.H. (1964). Trees and climatic history in Scotland. *Q. Jl. R. met. Soc.* 90, 382–94.

McVean, D.N. (1956). Ecology of *Alnus glutinosa* (L.). Gaertn. V. Notes on some British alder populations. *J. Ecol.* 44, 321–30.

McVean, D.N. (1958). Island vegetation of some West Highland fresh-water lochs. *Trans. Proc. bot. Soc. Edinb.* 37, 200–8.

McVean, D.N. (1961). Flora and vegetation of the islands of St. Kilda and North Rona in 1958. *J. Ecol.* 49, 39–54.

McVean, D.N. (1964). Woodland and scrub. In *The vegetation of Scotland* (ed. J.H. Burnett), pp. 144–67. Oliver and Boyd, Edinburgh.

McVean, D.N. and Ratcliffe, D.A. (1962). *Plant communities of the Scottish Highlands.* H.M.S.O., London.

Moar, N.T. (1969a). Two pollen diagrams from the Mainland Orkney Islands. *New Phytol.* 68, 201–8.

Moar, N.T. (1969b). A radiocarbon-dated pollen diagram from North-West Scotland. *New Phytol.* 68, 209–14.

Moar, N.T. (1969c). Late Weichselian and Flandrian pollen diagrams from South-West Scotland. *New Phytol.* 68, 433–67.

Newey, W.W. (1966). Pollen-analyses of Sub-Carse peats of the Forth Valley. *Trans. Inst. Br. Geogr.* 39, 53–9.

Newey, W.W. (1968). Pollen analyses from South-east Scotland. *Trans. Proc. bot. Soc. Edinb.* 40, 424–34.

Nichols, H. (1967). Vegetational change, shoreline displacement and the human factor in the late Quaternary history of south-west Scotland. *Trans. R. Soc. Edinb.* 67, 145–87.

O'Sullivan, P.E. (1973). Pollen analysis of mor humus layers from a native Scots pine ecosystem, interpreted with surface samples. *Oikos,* 24, 259–72.

O'Sullivan, P.E. (1974a). Two Flandrian pollen diagrams from the east-central Highlands of Scotland. *Pollen Spores,* 16, 35–57.

O'Sullivan, P.E. (1974b). Radiocarbon-dating and prehistoric forest clearance on Speyside (East-Central Highlands of Scotland). *Proc. prehist. Soc.* 40, 206–8.

Paton, J.A. (1965). *Census catalogue of British hepatics* (4th edn.). British Bryological Society, Ipswich.

Pears, N.V. (1968). Post-glacial tree lines of the Cairngorm Mountains, Scotland. *Trans. Proc. bot. Soc. Edinb.* 40, 361–94.

Pears, N.V. (1972). Interpretation problems in the study of tree-line fluctuations. In *Forest meteorology* (ed. J.A. Taylor), pp. 31–45, Cambrian News Ltd., Aberystwyth.

Pennington, W., Haworth, E.Y., Bonny, A.P. and Lishman, J.P. (1972). Lake sediments in northern Scotland. *Phil. Trans. R. Soc.* B **264**, 191–294.

Ratcliffe, D.A. (1968). An ecological account of Atlantic bryophytes in the British Isles. *New Phytol.* **67**, 365–439.

Rymer, L. (1974). The palaeoecology and historical ecology of the parish of North Knapdale, Argyllshire. Thesis, University of Cambridge.

Smith, A.G. and Pilcher, J.R. (1973). Radiocarbon-dates and the vegetational history of the British Isles. *New Phytol.* **72**, 903–14.

Spence, D.H.M. (1960). Studies on the vegetation of Shetland. III. Scrub in Shetland and in South Uist, Outer Hebrides. *J. Ecol.*, **48**, 73–95.

Steven, H.M. and Carlisle, A. (1959). *The native pinewoods of Scotland.* Oliver and Boyd, Edinburgh.

Tittensor, R.M. and Steele, R.C. (1971). Plant communities of the Loch Lomond oakwoods. *J. Ecol.* **59**, 561–82.

Vasari, Y. and Vasari, A. (1968). Late- and post-glacial macrophytic vegetation in the lochs of northern Scotland. *Acta bot. fenn.* **80**, 120 pp.

11

The early Quaternary landscape with consideration of neotectonic matters

CUCHLAINE A. M. KING

Abstract

The landforms of Britain at the beginning of the Quaternary reflect the position of the country at the trailing edge of a continental mass undergoing sea-floor spreading from the mid-Atlantic Ridge. There has been a long-continued tendency for basin subsidence in and around Britain, with compensating elevation of the stable blocks, such as the Pennines. Thus erosion surfaces occur in the uplands and deep sedimentary basins outline the British Isles, for example in the Irish Sea and North Sea. Warping is evident as a result of these earth movements, and there is some evidence of antecedent drainage, in the Weald for instance. Uplift of the upland areas in Plio-Pleistocene times may have helped to initiate the ice sheets that subsequently covered much of the country.

Introduction

The aim of this chapter is to summarize ideas concerning the nature of the landscape in the early Quaternary before the ice sheets enveloped much of the land surface of Britain. Landscape is a function of the action of processes on materials. The processes in turn are dependent on the morphology, the climate, and the materials. The materials are dependent on past processes and landscapes, as far as the solid rocks are concerned; and they reflect the currently-acting or recent processes, as far as the superficial deposits are concerned. A complex and intricate feedback relationship therefore exists between the morphology, the materials, and the processes. In an attempt to clarify the account the controls on the events that led up to the production of the early Quaternary landscape will be considered first, and then the landscape will itself be discussed.

The Tertiary events that are relevant can be subdivided into those connected with base level changes and those connected with climatic fluctuations, although again there is feedback between the two. The changes of base level can be subdivided into those related to variations in sea-level, the eustatic element, and those related to variations in land-level, the tectonic or isostatic element. Evidence for these two types of base level change will first be considered, and then evidence for climatic conditions will be mentioned. The evidence can best be sought in an analysis of the processes that would operate under the conditions of relief and climate dependent on these controls. The processes produce both erosional and constructional forms and materials. It is these processes that are next assessed, leading to the analysis of the morphology in the second section.

The elements of the morphology that are significant include the nature of the coastline of Britain at the start of the Quaternary period, the relief of the country in terms of both surfaces and slopes between them, and the drainage pattern which is closely connected to the relief. The analysis of the drainage pattern can provide valuable evidence concerning the development of the landscape.

These topics are all ones on which it is difficult to come to any definitive conclusions, partly owing to the removal of evidence through erosion. Some progress, however, has been made in elucidating the problems to be explored.

Tertiary controls and processes

During the last 15 years the development of ideas concerning plate tectonics has shown that the earth's crust is much more mobile than was thought. Britain is placed in a critical position from the point of view of the effect of oceanic rifting on the development of the land surface. The British Isles are situated close to the margin of an actively splitting ocean, forming a trailing edge coast. The forces associated with the widening of the north Atlantic Ocean are thought to have played an active part in the tectonic response of Britain to oceanic rifting, and they have also affected the eustatic level of the sea.

The crustal movements associated with the splitting of the north Atlantic started about 70 million years ago (Taylor and Smalley 1969), about the end of the Cretaceous period. The events of the subsequent Tertiary must be considered in an attempt to explain the landscape of Britain in the early Quaternary. The many geophysical studies that have been carried out over and around Britain have provided much useful evidence of structural conditions related to the processes of global tectonics, which have been and are still active in Britain and the adjacent sea areas. The existence of deep basins between areas of uplift has been a consistent feature of the structure of the British Isles and the adjacent seas. Examples of basins of subsidence include the two Minch basins, several basins in the Irish sea and adjacent land, such as the Cheshire basin, and the important North Sea basin (Hall and Smythe 1974). Differential movement between the basins and intervening positive areas has been active in some areas throughout Mesozoic and Cainozoic times, and is still continuing.

The vertical movements in and around Britain may be associated with action in the upper mantle, causing tilting. A current moving eastwards towards Britain may have caused differential tilting to start in Britain about 50 million years ago, when the pattern of mantle convection may have changed (Taylor and Smalley 1969). The British Isles could have been raised out of the sea and tilted down to the east as the currents were deflected on approaching the thicker continental crust. The subsidence of the North Sea could be related to the failure of the current to regain its original strength as its original direction was resumed. Uplift would occur in the high pressure zone to the west where the continental crust first thickened, with subsidence in the low pressure zone further east. The western uplift could have caused bending of the crust and graben formation, resulting in the initiation of the Irish Sea about 30 to 40 million years ago. Ireland now appears to be stable, but Britain is still tilting.

Sea-level changes—the eustatic factor

By far the most important element in the eustatic fluctuations of sea level in the Quaternary has been the glacio-eustatic effect, but other elements must also be considered especially in the period leading up to the Quaternary. The most significant of these are related to plate tectonics, and they include subsidence of the ocean crust, uplift of mid-ocean ridges, and changes in the length of the mid-ocean ridge system (Flemming and Roberts 1973). It is possible that the late Miocene uplift of the southern mid-Atlantic ridge could have resulted in a 10% reduction of ocean depth and an eustatic rise of about 500 m. This suggestion is supported by estimates of a middle Miocene transgression of 350 m. Maximum changes due to these causes range between a rise of 300 m and a fall of 800 to 1000 m. Fluctuations in the curve of eustatic sea-level are related also to changes in the rate of spreading from the oceanic ridges, subsidence occurring during accelerations and uplift during decelerations.

Tanner (1968) has assembled much of the data relevant to changes of sea-level during the Tertiary period, and he suggests a generally falling eustatic sea-level following the great Cretaceous transgression. He considers that during the late Cretaceous and early Tertiary, sea-level dropped about 50 m in 70 million years, at a mean rate of 0·7 mm/1000 years. During the Mio-Pliocene period there appears to have been a further fall of 75 m in 25 million years, a rate of 3 mm/1000 years. The increasing rate of fall in the later Tertiary he thought to be due to the growth of the Antarctic and Greenland ice sheets during this period. There is now evidence, however, that the Antarctic ice sheet has been in existence at least during the last 5 million years snd probably since the Miocene.

Drewry (1975) has suggested that the Antarctic ice-sheet was present 5·0 million years ago, while Le Masurier (1972) has put forward arguments for the presence of an ice sheet in Antarctica during nearly the whole of Tertiary time since the Eocene, so that its part in eustatic sea level changes must be considered with caution. It has been suggested that a major glacio-eustatic sea-level fall in the Eocene was related to the initial growth of the Antarctic ice sheet. A second major lowering probably took place in the early to middle Pliocene culminating in the Pleistocene when the northern hemisphere ice sheets developed. Glaciation in high latitudes of the northern hemisphere may have started 10 to 13 million years ago in south-east Alaska and 6 million years ago in the high Arctic. Data from the Arctic and Antarctic suggest that glaciation was extensive as now in the late Miocene and early Pliocene times. From middle Pliocene on, the northern hemisphere ice sheet was already as large as at present.

The acceleration of the eustatic fall of sea-level continued into the Pleistocene. West (1972) has suggested that a rough estimate of the eustatic fall of sea-level around the time of the Plio-Pleistocene boundary was of the order of 50 m. One of the significant features of the Pleistocene was the very rapid eustatic fluctuations of sea-level related to the fast growth and still quicker dissolution of the two great northern hemisphere land-based ice-sheets. These changes become better known and are seen to be very complex as the present is approached. They are superimposed in a complicated fashion on the changes in the elevation of the land masses as a result of warping and tectonics.

Mitchell (1972), in his discussion of the Pleistocene history of the Irish Sea, gives evidence that at the Coralline Crag/Red Crag transition, sea-level was at least 45 m above its present height in the Irish Sea. The evidence is based on the St. Erth Beds, which can be followed to a height of 35 m, and which were laid down in water at least 10 m deep. They have a southern fauna related to the Coralline Crag in its final phases, although the pollen suggests a lower Pleistocene age.

Land-level changes

Despite the apparent stability of Britain there is ample evidence that there have been major earth movements during the Tertiary and that these have played a very important part in the shaping of the landscape over which the glaciers advanced in the Pleistocene. One of the areas close to Britain where the evidence is particularly abundant is the North Sea. Clarke (1973) has shown that the North Sea has subsided steadily and very considerably during the Cainozoic. The maximum rate of subsidence increased from about 1 cm/1000 years during the Cretaceous to over 50 cm/1000 years during the Quaternary. Half of the subsidence has taken place since the Miocene during the last 15 million years. This subsidence is still affecting the southern part of the North Sea, and Quaternary deposits in the centre of the basin exceed 750 m in thickness.

Adjacent land masses have also undergone warping and faulting during the Tertiary, and a useful summary of relevant data has recently been presented by George (1974). He stresses the strength of tectonic activity in Britain during the Eocene and Oligocene, before the Miocene 'Alpine' earth movements. He considers that the landscape of Britain is geologically young. His analysis starts from the blanket cover of Chalk, which probably was very widespread but not uniform. The importance of the Chalk cover has probably been overstressed, and it seems likely that it was mutilated and extensively removed by early Eocene. Eocene strata were deposited on an emergent dome of Chalk at least in the south. The strata change from marine through estuarine to fluviatile to the west and north-west of the Hampshire Basin. The overstep westwards means that the Chalk had been entirely removed from Devon before the deposition of the

Eocene. In Wales and northern England Palaeozoic rocks were exposed to subaerial denudation throughout the Eocene. In Scotland and Ulster volcanic rocks of Eocene age also mainly rest on rocks older than the Chalk. The Chalk must have been removed from Ulster by the Oligocene, as there are no flints in beds of this age. While volcanic activity was taking place in northern Britain, in the south folding was severe, with amplitudes of more than 400 m in the London and Hampshire Basins.

Deformation continued throughout the Oligocene period, during which the Bovey Tracey beds were deposited. They rest unconformably on the Palaeozoic strata only 1·6 km from the Upper Greensand and Chalk outcrop. This relationship indicates sharp folding and faulting and a rugged terrain, giving rise to sand and gravel sedimentation. The Mochras borehole in Merioneth shows 550 m of Oligo-Miocene sediments resting discordantly on Lias, with no intervening younger Mesozoic rocks, including the Chalk (Woodland 1971). The Tertiary Mochras Beds rest directly on New Red Sandstone in Tremadoc Bay. Thus differential movement of the basins and blocks was active.

The movements continued on into the Miocene, and the pattern of a positive block over most of Britain surrounded by sinking basins persisted. These basins, such as the southern Irish Sea and Cardigan Bay, accumulated up to 7000 m of sediments in post-Carboniferous times and more than 600 m in the Cainozoic. In the Oligo–Miocene the Llanbedr fault on the land side of Cardigan Bay developed a throw of more than 800 m. Miocene faults and folds in Ulster are similar or larger than those in Hampshire; it is therefore, not possible to assume that little warping took place in northern Britain while the major earth movements were taking place in the south-eastern part of the country. It is likely that active folding also took place in the English Midlands, the Cheshire basin, the Solway area, and the Midland Valley of Scotland.

There is some evidence of sedimentation on the land during the Mio-Pliocene transition period. The Brassington formation, described by Walsh *et al.* (1972), is thought to be of Neogene age. It is considered to be a fluviatile deposit laid down by southward flowing rivers about 7 million years ago over a low gradient surface eroded across mainly Namurian strata. Since its formation relics of the formation have been preserved in deep sink holes that have progressively deepened to allow a total subsidence of 150 to 250 m. The sink holes have developed mainly in dolomitized limestone on the Carboniferous Limestone plateau of the Peak District in the Dove-Derwent interfluve area. The surface on which the Brassington formation was originally laid down near sea level has been raised by 450 m. The present upland surface, however, lies at 100 to 150 m below the 450 m level. This surface, often called the 1000 feet (300 m) erosion surface, formed during the Pliocene period, and has been modified during and after its formation by limestone solution. The surface was uplifted in late Pliocene or early Pleistocene times, when the dales that now dissect it were incised into the limestone.

Thick Pliocene deposits, descending to a depth of 200 m, occur in the Neogene succession of the southern Irish Sea, while strata of this age are 300 m thick off Norfolk, indicating continued pulsing subsidence of the basins. The very late Pliocene Coralline Crag over-steps from London Clay on to the Chalk, these sediments being younger than the Diestian Lenham Beds. Thus the range of altitude of the Pliocene strata of southern England between the land and the offshore zone is about 800 m, indicating continued crustal warping in the Neogene and later.

George (1974) concludes that the plateau form of upland Britain up to heights of 950 to 1000 m is a product of Neogene erosion, its elements descending in an unbroken geomorphological sequence to the coastal plateaux at 200 m and lower, which are dated as Pliocene in southern Britain. The degree of subsidence of the marginal basins is commensurate with the amount of uplift of upland Britain, indicating that the upland plains are not merely a result of eustasy. Block uplift has been dominant, but warping has also been active. Correlation by altitude is thus not possible except in narrow belts. The warping is still continuing according to the evidence given by Churchill (1965), who shows that in the last 6500 years the Liverpool Bay area has been uplifted 6 m, while the area between the Humber and the Thames has subsided by the same amount.

Evidence of climatic variation

The best evidence of climatic conditions is found by examining the nature of the superficial materials. Morphology can provide clues but may be misleading. The recognition of pediments in widely separated areas, including the Midlands and south-west by Dury (1972) indicates the danger of associating specific forms with special climatic conditions. Some of the pediments he recognized must have formed since the penultimate glaciation as they descend into Lake Harrison. They have continued to develop after solifluction that took place during the last glacial maximum.

The evidence of superficial material is more helpful, and Clark *et al.* (1967) discuss the origin and significance of the Sarsen stones of the Marlborough Downs from this point of view. The stones originated as duricrust formed under tropical weathering. The Sarsens are the remnants of Eocene strata laid down on attenuated Chalk only 107 m thick, a thickness that indicates that much erosion occurred before the deposition of the Eocene strata. The Sarsens are cemented with silica cement, which binds both silt and stones to form conglomerates in places. Large tabular blocks, 77% of which are 30 to 150 cm across, form stone trains. The blocks result from post-depositional induration below a vegetated surface and some contain roots. They probably formed on a planed surface under tropical conditions; the surface was most likely a post-Bartonian 'Eogene' surface cut in Oligo-Miocene times.

Evidence for a period of warm, moist climate is cited by Linton (1971) in his discussioñ of the low raised beach of southern Britain. The deep rotting of granite, conglomerates and slates before the cutting of the raised beach platform implies tropical conditions of warm and wet climate when sea level was also much lower. Unrotted parts of the rocks form tors, some of which have been drowned to form islands off the Breton coast. Linton considers that the warm, damp conditions occurred during an interglacial phase, which he argues was the last interglacial. Thus, if his time assessment is correct, it is necessary to exercise care in applying evidence for warm, damp conditions to an earlier phase of the Tertiary, or even the middle or late Tertiary.

The evidence given by superficial deposits
Superficial deposits, as already suggested, can provide evidence of climatic conditions, and they can also at times provide evidence of process. Marine deposits can be differentiated from fluviatile ones by their character and contained flora and fauna, while the deposits of down-wasting are also of considerable value in reconstructing landscape development.

Waters (1960) has discussed the significance of the superficial deposits on the well-developed upland plain of the Cretaceous plateau of east Devon and west Dorset. He argues that the character of the shingle found up to a height of 280 to 412 m on Staple Hill is such that the chatter-marked cobbles are of marine origin, thus indicating a sea level at this relative height which trimmed the surface on which the deposits rest. The surface is not horizontal, but falls gently south-eastwards, and the deposits predate the Miocene earth movements, so that Waters concludes that the marine gravels are early Tertiary in age. Their abrupt termination along a line running from the Black-down Hills to Axminster and Lyme Regis suggests a coastline at this position, as the deposits to the west of this line are fluviatile in character. The western part of the surface is flatter than the eastern, which is covered by the marine superficial deposits, while the western part has a covering of angular drift. It is possible that the western part was subaerially trimmed after the uplift in the mid-Tertiary earth movements, and thus the evidence of marine action is lacking. In the eastern area uplift also took place, and the marine deposits only survive near the stream sources, such as on Staple Hill. The post-folding, subaerial erosion was largely carried out on the early Tertiary deposits in the east where they were at a lower level. This material was removed by the planation to reveal the warped, marine-trimmed, sub-Eocene surface, which is cut largely across Cretaceous rocks. It was on the superficial deposits of this deformed surface that the major elements of the drainage pattern probably originated.

Clay-with-flints is an important and widespread superficial material in much of the area of the Chalk outcrop. The significance of this deposit has recently been reviewed by Hodgson *et al.* (1967, 1974) in relation to the associated soil horizons on the South Downs. It is a reddish, highly tenacious clay that immediately overlies the chalk; it contains nodular flints and is called Clay-with-flints (*sensu stricto*). Wooldridge and Linton (1955) assume that chalk solution would produce Clay-with-flints, but it has too much clay for it to have been formed only from chalk. The presence of quartz pebbles, sandstone, and fragments of ironstone, as well as Sarsens, indicates the former existence of a superficial layer of Tertiary material.

Four main processes produced the Clay-with-flints from the Reading Beds, which formed a thin, broken cover over the Chalk. The first process was weathering of the clay fraction; the second the downward movement of clay and its deposition lower in the profile; thirdly chalk dissolution occurred below; and fourthly a mixture of flints, chalk residue, and the overlying layer was formed by cryoturbation in a cold period. The development of Clay-with-flints was a cyclical process, with decalcification, weathering, and translocation taking place in warmer phases. The important point is that Clay-with-flints is not an uncontaminated chalk residue. The Plateau Drifts of the Chiltern Hills suggest a more complex development than the Clay-with-flints of Sussex.

Another possible interpretation of some of the superficial deposits of southern England has been advanced by Kellaway *et al.* (1975). They suggest that the whole country was overrun by Saale ice, which left morainic material near Selsey Bill and fluvioglacial sediments in the Hampshire New Forest, forming the upper Plateau Gravels above 128 m, and near Slindon and Brighton on the south coast. The latter deposits have usually been interpreted as a raised beach. They also suggest that the Clay-with-flints is a degraded and piped till in the North Downs, containing relict material from the Eocene strata and being younger than Red Crag (Waltonian). In support of their views they point to the presence of erratics, which occur along the south coast as far east as Pagham Harbour, Brighton, and Seaford Head, while Scandinavian material has been found at Purley in Surrey. Much further and more conclusive evidence is needed before these views can be accepted.

The early Quaternary landscape

The coastline

Shotton (1962) assembled the evidence then available concerning the physical character of Britain in the Pleistocene, including the land connection with the continent and Ireland. He reached the conclusion that the Straits of Dover were not open in the early Pleistocene. There is evidence that the Lenham Beds of Diestian age, now at about 200 m above sea level on the North Downs of Kent, were laid down in water about 72 m deep. Subsequent uplift has, therefore, been of the order of 275 m. At this time, therefore, there is likely to have been a water connection through the site of the Straits of Dover, allowing warm fauna from the Mediterranean to be included in the Lenham Beds in the early Pliocene. The next datable deposits are the Netley Heath Beds of the lowest Pleistocene, the base of the Red Crag. It is a shallow-water deposit, and now occurs at about 190 m. Thus a 76 m uplift must have taken place in the North Downs during the later part of the Pliocene. This uplift could have caused the emergence of a land bridge of chalk. Further uplift after the deposition of the Netley Heath Beds must have taken place in the Lower and Middle Pleistocene, as the Netley Heath Beds are now at 190 m. Beds of a similar age occur at 120 m at Rothamstead in the Chiltern Hills, but are less than 30 m at Walton-on-the-Naze. At Ludham in Norfolk Upper Red Crag shelly sands descend to 50 m below sea level. Most of the uplift probably took place in the early Pleistocene, and by the time of the Norwich Crag and Cromer Forest Bed Series the Straits of Dover were certainly closed, and a land connection with France existed probably until fairly late glacial times. The biological arguments also strongly support the idea of a land bridge with the continent during much of the Pleistocene, allowing recolonization to take place in interglacial periods.

　　Destombes *et al.* (1975) indicate a chalk ridge across the site of the Straits of Dover in the pre-Saale period, allowing a land connection between England and France in the Hoxnian interglacial. They suggest that this ridge marked the eastern limit of an ice sheet advancing from the west up the English Channel, and that the ridge was subsequently overrun and breached by late Saale ice moving from the north. They consider that an 18 km-long trench running west 10° north 10 km south of Dover could be a tunnel valley cut by meltwater from the early English Channel ice. The feature, however, may be much younger on palynological evidence, and could well be tectonic in origin, or formed by tidal streams. The evidence for an English Channel ice-sheet is by no means convincing or conclusive.

Funnell (1972) has discussed the history of the North Sea during the Tertiary and Quaternary. The Palaeocene sea, he considers, was about its present extent, but relatively cool. During the Eocene, however, sub-tropical conditions occurred, and muddy sediments were laid down in the London and Hampshire basins in the south-west, where a strait connected with the Atlantic. This was closed after the middle Oligocene, when earth movements started in the Wealden area, resulting in a cooler North Sea in the early Miocene. A maximum warm temperate state probably occurred in mid-Miocene, when the connection with the Atlantic may have been re-established, with similar conditions carrying on into the Pliocene. After Red Crag times the southern route of access was closed, and any warm fauna must have migrated around the north of Britain, for example the few specimens from the Ludham Crag, which is correlated with the Upper Red Crag.

A land connection between Ireland and the rest of Britain seems less likely, according to Shotton (1962) in the early Pleistocene. The depth of the Irish Sea is such that a land connection during earlier interglacials was unlikely, and earth movements are also unlikely to have created a link. The virtual absence of fossil mammalian fauna in Ireland supports the absence of a land bridge.

The present outline of Britain probably developed during the Tertiary as the movement between the surrounding subsiding basins and the inland areas of uplift became marked. This differential movement accounts for the contrast between the relative smoothness of the offshore relief, and the greater relief of the inland zone, as uplift caused incision and rejuvenation. The repeated passage of the sea across the inner continental shelf during the rises and falls of Pleistocene sea-level, with the waxing and waning of the ice sheets, has increased the smoothness of the offshore relief. The hard-rock cliffs of the exposed western coast, separating offshore and inshore areas, show evidence of long continued activity by both marine and subaerial forces. Many of these cliffs are probably of considerable antiquity and composite in character. The contrast between cliffs and marine planation surfaces thus forms a striking element of present relief.

Relief and drainage

Recent work has led to a re-interpretation of the Tertiary history of Britain and hence to a different appraisal of the landscape at the beginning of the Pleistocene. It is now realized that the land is much more mobile, in recognition of the position of Britain at the trailing edge of a continental block undergoing sea-floor spreading from the mid-Atlantic Ridge. There is ample evidence that uplift has been active for a long time in many parts of Britain, especially in the uplands, such as the Pennines, while at the same time deep basins of sedimentation have been filled in the marginal seas, such as the Irish Sea and the North Sea. Thus the general outline of Britain has been sketched out for the major part of the Tertiary period, although the island status of Britain has varied throughout the period, and the sea has transgressed across the land at intervals to varying extents after the extensive Chalk transgression of the Cretaceous. The importance of the Eocene transgression has been increasingly appreciated, largely through a study of the superficial deposits that mantle much of the southern part of England.

The sub-Eocene surface has been recognized as forming an important part of the landscape in the southern part of England, which is generally regarded as unglaciated. The surface is apparent from the Clay-with-flints and Sarsen stones which are the weathered and consolidated relics of early Tertiary deposits mixed with the residue of Chalk solution. The surface has, however, been considerably warped by the earth-movements that culminated in the middle Tertiary, but which have continued both before and since this period. Drainage probably developed on the sub-Eocene surface, and in places it seems likely that the original pattern has continued to run across the rising folds, thus showing antecedence, while elsewhere streams have become adjusted to the major structures. In only a few places is there evidence of discordant drainage due to superimposition from the deposits of the most recent transgression in the Calabrian.

The Calabrian transgression in the earliest part of the Pleistocene was an important event in the evolution of the landscape of southeast England, and its influence on the drainage of the area has been studied by Jones (1974). He makes out a case for antecedence in relation to the effects of the Calabrian transgression. There is only good evidence for the transgression in the London Basin, where there is a bench with early Pleistocene marine sediments on it. Outside the London Basin the evidence adduced for the transgression is usually the discordant nature of the drainage, but where the morphological evidence is sound, discordance is rare and vice versa.

There is no evidence that marine erosion resulted in a discordant drainage pattern, and the Calabrian Sea probably advanced over a surface of considerable relief of at least 200 m, into the London and Hampshire Basins and the clay vales of the Weald.

There must have been north to south drainage across the Chalk of the South Downs before the transgression. A marine trimmed surface 20 to 65 km wide would have been needed for the Sussex and Hampshire basins respectively to account for the drainage discordance as suggested by Wooldridge and Linton, and large volumes of strata would have had to be removed. Although the Calabrian Sea probably covered large parts of Sussex and Hampshire, it probably was impotent and did not affect the drainage pattern.

Antecedence can account for the drainage pattern across the secondary folds if the rivers flowed southwards in the late Tertiary, as appears likely. The original pattern was of major east-flowing streams the Thames and Solent, with lateral tributaries. The concordant pattern of the North Downs is similar to the discordant pattern of the southern Weald, and a common origin could explain the similarity. In the north tectonic effects were slight, but in the south numerous periclines formed. This tectonic disturbance started in the lower Cretaceous and continued through to the early or middle Pliocene, with warping going on until the present. Growth of secondary structures was probably prolonged, and therefore slow, allowing antecedence to develop. The scale of discordant relationships is generally small in Wessex; folds cut through have amplitudes of less than 180 m, and the rivers had more than 2 million years to erode through them in the Hampshire basin. The main rivers of Sussex are concordant with the major structures while the headwater streams in the Weald in general show more concordance with minor structures than the larger streams in Sussex. The folds may have been increasing in amplitude since they were formed by pressure release processes. The rivers of southern Britain probably resumed their former courses as the Calabrian Sea withdrew.

In the northern part of Britain and those parts of the country that were covered by the ice-sheets there is very little evidence in the form of deposits to help elucidate the events of Tertiary time, and thus to reconstruct the conditions at the beginning of the Pleistocene. Nevertheless the morphological evidence does provide some clues, and a few small patches of Neogene deposits have survived, including the Brassington formation in the sink holes of the Derbyshire Dome area. Planation surfaces are conspicuous features in parts of the Midlands and northern Britain, those of the Pennines being especially well developed. The 300 m summit surface of the southern Pennines is a conspicuous feature, while further north in the Askrigg and Alston Blocks of the central Pennines, which is a relatively stable area, a gently warped summit surface is well developed. This surface is probably of late Pliocene or early Pleistocene age, and it indicates that uplift has continued throughout the Pleistocene, as the surface is now at a considerable altitude, reaching at least 600 m in places. The elevation of this surface is probably partly due to long continued tectonic upwarping, resulting from granite intrusion. The altitude of the surface was also increased as a result of the growth of the northern hemisphere land-based ice sheets towards the end of the Pliocene and the beginning of the Pleistocene, causing an accelerating eustatic fall of sea level.

It is probable that in stable areas sea-level has fallen eustatically by about 200 m since the Calabrian transgression, as a surface and deposits at this level have been widely recognized; however, warping is also implied as the deposits occur at different levels. The time of the Calabrian transgression was probably about 2 million years ago (Kidson 1968). As the Calabrian sea was transgressive there is no reason why the platforms below it, such as the 130 m (430 feet) one, were not cut before it. Its relative impotence has already been noted, so that there is little reason why it should have obliterated previously cut surfaces at lower levels.

Conclusions

At the opening of the Pleistocene the Calabrian Sea had transgressed in a fairly quiescent way to about 200 m above the present sea level, and in the upland parts of the country, above the reach of the sea, subaerial erosion had probably produced a fairly subdued landscape. In the 2 million years since the sea withdrew eustatically, falling sea level and local warping have enabled subaerial agencies to produce the present landscape. In the upwarped areas there are dissected upland surfaces, many of which have been further modified by glacial and periglacial processes. In the downwarped areas, notably adjacent to the southern North Sea, the land has received further marine sediments, the Upper Crags, as well as later glacial deposits. There was a long period of subaerial modification of the landscape prior to the Anglian period during which uplift caused rejuvenation, leading to increased relative relief. This greater elevation must have been one factor in the initiation of the local ice-sheets, which first reached the Midlands and south during the Anglian period. Thus the landscape has increased in relief throughout the Pleistocene, and the ice sheets of the later Pleistocene have done much to enhance the interest and beauty of the landscape of Britain.

References

Churchill, D.M. (1965). The displacement of deposits formed at sea level 6500 years ago in southern Britain. *Quaternaria* 7, 239–49.

Clark, M.J., Lewin, J. and Small, R.J. (1967). The Sarsen stones of the Marlborough Downs and their geomorphological implications. *Southamp. Res. Ser. Geogr.* 4, 3–40.

Clarke, R.H. (1973). Cainozoic subsidence in the North Sea. *Earth planet Sci. Lett.* 18 (2), 329–32.

Destombes, J-P., Shepard-Thorn, E.R. and Redding, J.H. (1975). A buried valley system in the Strait of Dover. *Phil. Trans. R. Soc.* A 279, 243–56.

Drewry, D.J. (1975). Initiation and growth of the East Antarctic ice sheet. *J. Geol. Soc. Lond.* 131, (3), 255–73.

Dury, G.H. (1972). A partial definition of the term 'pediment' with field tests in humid climate areas of Southern England. *Trans. Inst. Br. Geogr.* 57, 139–52.

Flemming, N.C. and Roberts, D.G. (1973). Tectono-eustatic changes in sea level and seafloor spreading. *Nature, Lond.* 243, (5401), 19–22.

Funnell, B.M. (1972). The history of the North Sea. *Bull. Geol. Soc. Norfolk* 21, 2–10.

George, T.N. (1974). Prologue to a geomorphology of Britain. In *Progress in geomorphology: papers in honour of D.L. Linton.* (Ed. E.H. Brown and R.S. Waters) pp. 113–125. British Geomorphological Research Group.

Hall, J. and Smythe, D.K. (1974). Discussion of the relation of Palaeogene ridge and basin structure of Britain and the north Atlantic. *Earth planet. Sci. Lett.* 19, 54–60.

Hodgson, J.M., Catt, J.A. and Weir, A.H. (1967). The origin and development of Clay-with-flints and associated soil horizons on the South Downs. *J. Soil Sci.* 18, 85–102.

Hodgson, J.M., Rayner, J.H. and Catt, J.A. (1974). The geomorphological significance of Clay-with-flints on the South Downs. *Trans. Inst. Br. Geogr.* 61, 119–29.

Jones, D.K.C. (1974). The influence of the Calabrian transgression on the drainage evolution of south-east England. In *Progress in Geomorphology: papers in honour of D.L. Linton.* (ed. E.H. Brown and R.S. Waters), pp. 139–158. British Geomorphological Research Group.

Kellaway, G.A., Redding, J.H., Shephard-Thorn, E.R. and Destombes, J.-P. (1975). The Quaternary history of the English Channel. *Phil. Trans. R. Soc.* A 279, 189–218.

Kidson, C. (1968). The role of the sea in the evolution of the British Landscape. In *Geography at Aberystwyth.* (ed. E.G. Bowen, H. Carter and J.A. Taylor), pp. 1–21. University of Exeter.

Le Masurier, W.E. (1972). Volcanic record of Antarctic glacial history: implications with regard to Cenozoic sea levels. *Polar geomorphology.* (ed R.J. Price and D.E. Sugden), *Inst. Brit. Geog. Spec. Publs.* 4, 59–74.

Linton, D.L. (1971). The low raised beach of southern Ireland, South Wales, Cornwall, Devon and Brittany and its relations to earlier weathering and later gelifluction. *Quaternaria.* 15, 91–8.

Mitchell, G.F. (1972). The Pleistocene history of the Irish Sea, second approximation. *Scient. Proc. R. Dubl. Soc.* A 4 (13), 181-99.

Shotton, F.W. (1962). The physical background of Britain in the Pleistocene. *Advmt. Sci.* 19 (79), 193-206.

Tanner, W.F. (1968). Tertiary sea level symposium—introduction and multiple influences on sea level changes in the Tertiary. *Palaegeog. Palaeoclimatol. Palaeoecol.* 5, 7-14, 165-71.

Taylor, R.L.S. and Smalley, I.J. (1969). Why Britain tilts. *Sci. Jl.* July 1975, 54-9.

Walsh, P.T., Boulder, M.C. Ijtaba, M. and Urbani, D.M. (1972). The preservation of the Neogene Brassington formation of the southern Pennines and its bearing on the elevation of Upland Britain. *J. Geol. Soc. Lond.* 128, 519-59.

Waters, R.S. (1960). The bearing of superficial deposits on the age and origin of the upland plain of east Devon, west Dorset and south Somerset. *Trans. Inst. Br. Geogr.* 28, 89-95.

West, R.G. (1972). Relative land sea level changes in south-east England during the Pleistocene. In: A discussion on problems associated with the subsidence of south-east England. *Phil. Trans. R. Soc.* A 272 (1221), 87-98.

Woodland, A. (ed.) (1971). The Llanbedr (Mochras Farm) borehole. *Rep. Inst. Geol. Sci.* 71/18.

Wooldridge, S.W. and Linton, D.L. (1955). *Structure, surface and drainage in south-east England.* 2nd Edn. (1st Edn Inst. Brit. Geog. 10, 1938). Philip, London.

Wright, A.E. and Moseley, F. (ed.) (1975). *Ice Ages: Ancient and Modern. Geol. J. Spec. Issue* 6, Liverpool.

12

River terraces

K. M. CLAYTON

Abstract

The terraces of the major rivers in England are described; some may be dated by reference to glacial history or included fossils. In the past the emphasis has been on control by sea-level changes, but correlation between river basins shows that the amount of dissection has varied tremendously, so that the height of terraces above the modern flood-plain is neither a guide to their age, nor to past sea-level. A climatic interpretation of terraces must face the fact that several include interglacial material, while in many cases the modern channel is still reworking the flood-plain deposits, despite the current interglacial conditions.

General introduction

Although the river terraces of the British Isles have long been studied, knowledge of them remains incomplete. For most of its history, the official Geological Survey has neglected Quaternary deposits, and the efforts of individual research workers have inevitably led to a preponderance of local studies with awkward gaps between them. However, more serious than this patchy cover is the wide variety of concepts about the nature of river terraces held by those who have surveyed them. Much of the work has been based on the assumption that they represent aggradation in response to high (interglacial) sea-levels, and it has frequently been suggested that the relative height of the terrace flat above the modern floodplain is an indication of the height of the related sea-level. In addition the classic work of French archaeologists led to the enthusiastic adoption of included palaeo-liths as a basis for dating and correlating terrace gravels. Attempts to correlate river terraces between the major basins on either of these concepts failed, and this has discouraged the erection of a general scheme or model of terrace development and chronology. Today many authors follow the continental concept of climate control of terrace formation, but it is significant that so far no general correlations have been put forward on this basis.

It will come as a surprise to some overseas visitors to learn that the climatic approach to river terrace development is treated with caution by many British geologists and geomorphologists. In part this is because language problems have kept us insulated from much of the continental literature, although Zeuner (1945) introduced the concept of climatic control thirty years ago. In a small island it is always easier to envisage control over geomorphological development by the sea, and even Zeuner matched his concept of cold, aggraded terraces upstream, with interglacial high sea-level terraces (thalassostatic terraces) in the lower part of each river basin. Before Zeuner, almost all British work assumed sea-level control throughout the river basin, and the concept has survived into much post-war work as well. However, appreciation of the complexities of climate / river regime interaction (Schumm 1965) has reduced the assurance of this approach.

Much of this chapter describes the terrace systems of the main river basins, the Thames, Trent, and Severn/Avon. Nevertheless, it is helpful to discuss some of the conceptual problems presented by river terrace formation, and to examine the pitfalls of some of the assumptions that have guided past interpretations. It is impossible to describe field evidence without a whole series of assumptions, few of which are spelt out in the original text, often because they are thought at the time to be so obvious. It is only when inconsistencies arise that we have to re-examine the assumptions as well as the evidence to see where we went wrong.

One way out of a number of difficulties over terrace sequence has been the concept that we may somehow separate the internal stratigraphy (the geology) of terraces from the height sequence of the terrace flats or surfaces (their geomorphology). This separation gives a freedom to explain geological sequences within terrace gravels as shown by fossils or granulometry without the constraint of the time sequence required by the stairway of terrace surfaces. Thus it is possible, as indeed Shotton (1953) has suggested in the Avon, to postulate aggradation giving a gravel thickness more than the difference in height between two terraces. If the river later cuts a lower terrace flat within this thick aggraded sequence, the upper terrace is underlain by younger gravels than the lower, reversing the usual arrangement. At first sight this is a very likely sequence of events, yet as a fluvial process, the cutting of the lower terrace flat without any reworking of the underlying gravel at the base of the aggraded sequence seems unlikely. It is of course entirely possible for the older gravels of thick aggraded sequences to survive beneath a higher terrace which marks the completion of the aggradation, but the gravels underlying a lower terrace flat must at the very least have been reworked to the full channel depth in flood, and thus are the same age as the terrace flat.

Not all the elements of the gravel associated with a terrace flat will date from the time the flat was formed. Archaeologists have long accepted the reworking of artifacts, and derived fossils of natural or human origin are rather common in river deposits. Further, abandoned channels, whether braided or meandering, will gradually fill up with younger deposits which eventually build up to the general terrace level; dates from these are younger than the period of terrace formation.

There is a repeated tendency in the literature to assume that the coarse gravels typical of most British river terraces are very different from the material moved in the modern channel. In part this has been guided by images of a glacial past, with huge meltwater discharges combining with bare slopes to produce a discharge/load regime wholly different from the contemporary situation. The concept is further supported by the sediment normally seen in or near the modern river bed; sand silts and mud at low water, and sandy or silty overbank deposits. Evidence on modern channel migration both supports and conflicts with this concept. The Thames channel has not moved far from its present position over the last few thousand years, and it does seem that contemporary flood discharges are unable to shift the calibre of load that lies beneath the present floodplain. On the other hand, the River Dove in Derbyshire is meandering so freely across its floodplain that it has already moved far from the county boundary surveyed along the channel just over a century ago. The main changes in peak discharge may have occurred where permafrost sealed porous outcrops (such as the chalk), and the Thames would be likely to be strongly affected in this way, perhaps explaining why it seems to be trapped in its floodplain gravels.

A recurring concept in the formation of terraces, whether by climatic control or by high sea-levels, is aggradation. In the upper basin, so conventional wisdom tells us, the high sediment contribution from slopes in periglacial conditions of glacial times dominated the regime and caused aggradation. Aggradation may also result from braided rivers, especially where heavily laden meltwater from a glacier is involved. Such aggraded surfaces are dissected in the subsequent warmer period when a vegetation cover reduces sediment discharge, causing the river to erode and incise. In the lower course, while a falling sea-level causes incision, aggradation seems the inevitable result of a high sea-level (Fig. 12.1). However, such aggradation is unlikely to have extended far upstream, as studies on channel response upstream of reservoirs and millponds show (see Leopold, Wolman and Miller 1956).

It will be suggested here that many terraces are not formed by aggradation of braided rivers, but by cut and fill by meandering channels. This is not to say that there have not been regime changes, for this is the most probable cause of a terrace sequence. Climatic change is likely to have been the main cause of these regime changes, although we cannot always rule out tectonic uplift, and river diversion has also been a factor. Some past floodplains will have been formed by braided streams, but by including cut and fill by a meandering river as a floodplain-forming process, we are free from the rigid constraints of climate control.

This is not the point to discuss at length the possibility that most British river terraces are the dissected cut and fill floodplains of streams not unlike those forming floodplains today. Nevertheless, we may note the consistent thickness of terrace gravels of each river, more readily explained as controlled by the depth of flood scour than as a repeated consistent depth of aggradation by a braided stream. We may also note the common slope of terraces, although this may be controlled as much by the recurrence of similar size reworked sediment in the coarser grades of each terrace as by the channel style at the time of deposition. However, it is noticeable that

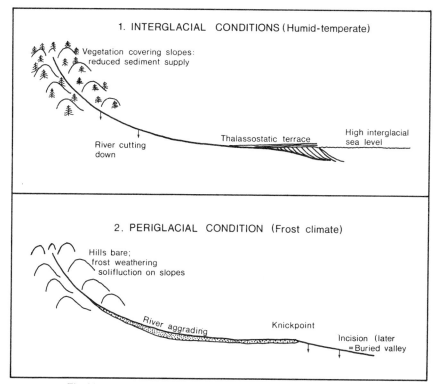

Fig. 12.1. Conventional view of climatic control of river behaviour.

the one terrace of the River Thames thought to have been aggraded in response to glacial blocking of the valley has an abnormally low slope compared with all the other terraces (Hare 1947), which slope at much the same angle as the modern floodplain. Finally, we may also note that a number of British river terraces are demonstrably not glacial; as is shown by the interglacial age of the Beeston terrace of the Trent (Jones and Stanley 1974), the Boyn Hill (at Swanscombe, King and Oakley 1936), and Upper Floodplain (e.g. Trafalgar Square, Franks 1960) terraces of the Thames.

At each stage marked by a terrace, the river was flowing to a former sea-level, so that each terrace necessarily leads downvalley to a former sea-level. In fact the downstream marine equivalents of these former floodplains are rarely preserved: many will have lain far beyond the present coast, for only in interglacial times did the coastline lie near the present coast of the British Isles. Since the position of the river mouth is essential if we are to establish former sea-level (see for example, the attempts to determine former sea-levels discussed by Miller 1939), the terraces as such, contrary to common

Table 12.1.
*Gradients of floodplain and principal terraces
(lower 100 km of mainstream above tidal limit)*

		gradient (m/km)
Thames:	Floodplain	0·41
	Lynch Hill terrace	0·41
	Boyn Hill terrace	0·38
Trent:	Floodplain	0·44
	Hilton terrace	0·47
Severn:	Floodplain	0·33
	Worcester terrace	0·48
	Main terrace	0·57

belief, tell us nothing at all about former sea-levels. Only where there are associated raised beaches or where estuarine or marine sediments are intercalated with the river deposits can we establish former sea-levels.

This account is limited to the main river basins of southern Britain, for only here are reasonably full sequences developed with at least the outline of adequate stratigraphical control. Quite full flights of terraces have been described from a number of southern rivers, especially those beyond the glacial margin. However, these tend to be even less well known stratigraphically than the more extensive terraces of the larger rivers, and in so far as a chronology has been suggested, it is derived by analogy (often on the basis of height above floodplain) with the sequence in the Thames basin. Further north complex terrace sequences are known but these are almost entirely related to the melting back of the last ice-sheet and they cannot readily be correlated from one basin to another. In Scotland the isostatic adjustments related to the growth and decline of the ice-sheet, coupled with major regime changes as the ice waxed and waned and discharges of sediment fell and rose again, have built some of the most perfectly geometrical of our terraces.

The Thames basin
The Thames terraces have been known in outline for a very long time; the names Boyn Hill and Taplow were used by Whitaker in 1889. Particular areas have been subject to far more detailed work, but this has never been extended through the basin on a systematic basis, and there are still two considerable major gaps in our knowledge. The

Middle Thames and the Upper Thames around Oxford are not satisfactorily correlated because of the scarcity of terrace remnants in the
Goring Gap between them. Downstream there is insufficient information in London itself, an area sadly neglected by geomorphologists
despite the frequent exposures caused by rebuilding.

The first account of the geology of the London Basin (Whitaker
1889) recognized three terraces above the present floodplain, the
Boyn Hill (100 foot = 32 m), Taplow (50 foot = 16 m), and Upper
Floodplain (25 foot = 8 m) terraces. Higher gravels were called 'Plateau gravels' and shown in pink on the map, and these included
everything from late-Tertiary gravels to Pleistocene river terraces that
happened to lie more than 100 feet above the floodplain. Work by
Wooldridge (1938) in particular established a series of terraces and
higher less well-preserved features he called 'gravel trains' (cf.
Deckenschotter) which were related to a series of older courses of
the Thames, disturbed on more than one occasion by glaciation,
when ice diverted the Thames south of the synclinal axis of the
London Basin and into its present valley. The highest of these gravel
trains was related to the earliest ice to reach the London basin, the
Anglian, and the gravel surface lay about about 100 m above the
Thames floodplain west of London. In 1947 Hare described eight
terraces west of London, dividing Wooldridge's lower gravel train
into the Harefield, Winter Hill, and Black Park terraces, but retaining
the names used by the Geological Survey for three of the terraces.

Some of these complications had been suggested at the time of the
now-classic synthesis by King and Oakley in 1936. This account
rested on the recognition of periglacial deposits which could be
traced as trails of Coombe rock (from the Chalk) and similar 'head'
deposits leading down into the relatively deep valleys formed during
successive glacial periods with low sea-level. The terrace gravels were
seen as aggraded fills in response to higher sea-levels. Much of their
correlation rested on palaeoliths and closely followed the approach
of Breuil in France. King and Oakley also noted the deep buried
channel of the Lower Thames, concluding that it was cut immediately
after the deposition of the Upper Floodplain terrace.

In recent years a number of fossiliferous sites have been investigated in the Thames valley and some of the more important ones
may be noted here. Further excavations at Swanscombe (Waechter,
Newcomer, and Conway 1970) have confirmed the Hoxnian age of
the gravels of the Boyn Hill terrace and their correlation with the
gravels lying near sea-level at Clacton. It is likely that the latter,
which lie some 70 km from Swanscombe, have been downwarped to
some extent as part of the North Sea basin movement. The old

concept, championed by Zeuner (1961) and Wooldridge (1958), that the elevation of the Boyn Hill terrace at Swanscombe records the Hoxnian sea-level (+32 m) is untenable, for downwarped or not, the gravels at Clacton are fluviatile and lie downstream and thus necessarily lower than those at Swanscombe. Indeed the overall slope from Swanscombe to Clacton is so close to the terrace slope above London that were there not other reasons for inferring downwarping of the North Sea basin, the idea could not be sustained on the evidence of the terraces alone.

A number of sites have been interpreted as Ipswichian: Trafalgar Square, Ilford, and Aveley. There is some dispute about how far all these fall into a single interglacial, and they certainly overlap deposits previously assigned to the Upper Floodplain and Taplow terraces. The Taplow above (and within) London is generally considered to show a cold environment with mammalian fossils from the type site near Maidenhead and plants from Endsleigh Gardens, so the warmer deposits of the Ilford–Aveley area are no longer called Taplow. The Aveley deposits in particular suggest estuarine conditions and if so lie near to the contemporary sea-level—in local terms suggesting an altitude of about 10 m O.D., although of course the stability of the area is unknown (West, Lambert, and Sparks 1964; West 1969).

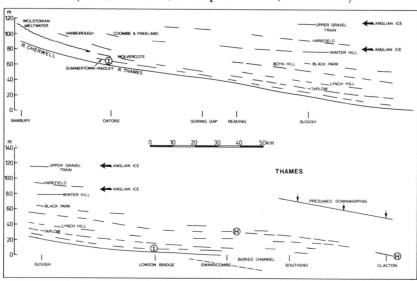

Fig. 12.2. Terraces of the River Thames. H = Hoxnian, I = Ipswichian

The problem in any physical linking of the terraces of the Oxford region with the Middle Thames has already been suggested. As the Cherwell carried overflow from Lake Harrison ponded by the

Wolstonian ice (Bishop 1957), we have a chance of linking that stage with the Upper Thames terraces. The terraces around Oxford include the Summertown–Radley (8 m above alluvium) with an Ipswichian fauna, and the Wolvercote (16 m) and Hanborough (32 m) terraces. In the Cherwell, the Wolvercote with its incursion of flints marks the Lake Harrison (Wolstonian) overflow (Bishop 1957), and is thus Wolstonian. It used to be thought that the Hanborough was Hoxnian, but Briggs and Gilbertson (1973) have shown that the sedimentology, fauna, and included periglacial structures indicate a cold climate, so it is also Wolstonian. Downstream the Hanborough must link with the Lynch Hill and the Wolvercote with the Taplow; in each case the fall of about 0·4 m per km is close to the gradient of the modern floodplain.

The Trent basin

We turn next to the Trent as it is a good example of the more restricted range of terrace development found within the glaciated part of the British Isles. In addition, its lower course turns northwards, so that it was diverted from its course during the glaciations, and some of the terraces diverge from the modern line of the river. Finally, it shows far more clearly than the Thames the extent to which the most recent significant rejuvenation has worked back to produce the youngest terrace downstream of Nottingham.

The highest terraces of the Trent have been called the Hilton terraces (Clayton 1953), and most workers have accepted them as true river terraces. They date from the melting of the Wolstonian ice-sheet, contain a high proportion of erratic material (> 70% Bunter quartzite pebbles) and lumps of till, and show disturbed bedding (Posnansky 1960). No traces of the Hilton terrace survive between Nottingham and Newark in the relatively narrow trench which marks an ice-marginal route of the Trent, cut obliquely across the Keuper dip-slope between Nottingham and Newark. However, between Newark and Lincoln very extensive patches of gravel cap the hills at Graffoe and Eagle at 25–30 m OD. These indicate that at this stage the Trent passed through the Lincoln gap, and the gravels continue southeast-wards alongside the River Witham. The course is likely to mark the decay of the Wolstonian ice-sheet, with a lobe of thick, relatively clean, ice surviving in the Vale of Belvoir and a further lobe blocking the present lower Trent valley in the Gainsborough area.

The best developed terrace of the Trent is the Beeston terrace (Swinnerton 1937). This is widely developed in the various valleys above Nottingham, sloping a little less steeply than the modern floodplain, so that it lies 3 m above the floodplain at Rugeley and

6 m at Beeston itself. Downstream there are again few remnants between Nottingham and Newark, but the terrace resumes as a well-
preserved gravel sheet leading towards the Lincoln gap and so presumably down the Witham. At Allerton, near Derby, *Hippopotamus*
was found in 1895 (Bemrose and Deeley 1896), suggesting an Ipswichian date, and recent work by Jones and Stanley (1974) has
confirmed this.

Most of the Trent valley escaped glaciation during the Devensian,
but ice entered the upstream parts of the basin, and may have pushed
into the Gainsborough area. At Long Eaton above Nottingham there
is a knickpoint, and for the remaining 50 km or so to the tidal limit,
the modern floodplain lies several metres below the floodplain terrace, which is of Devensian age and was fed upstream by outwash
from the Blithe and the Churnet. Below Long Eaton the Devensian
terrace is preserved by incision, but above the knickpoint, the contemporary channel is reworking Devensian material and so incorpor-
ating postglacial fossils into the older sediments. It is likely that the
rejuvenation of the Lower Trent is linked with (and may be caused
by) the diversion (perhaps by overflow rather than capture) of the
Lower Trent from the Lincoln gap to the Humber.

The Severn and Avon basins

We turn finally to the Severn, and its relatively well-known tributary,
the Avon. In both of these basins the changes brought about during
the later part of the Quaternary have been profound. The Avon is
no older than the close of the Wolstonian glaciation: prior to that the
area around Stratford-on-Avon drained north-eastward to the Soar
and so to the Trent (Shotton 1953). The Severn is even younger, for
the connection with the Shropshire lowland through the Ironbridge
gorge (Wills 1924) was initiated during the Devensian glaciation:
before then the Welsh Severn must have joined the Welsh Dee and
flowed northwards into Liverpool Bay.

The effect of these changes is to have greatly increased the area
that drains to the Bristol Channel and thus to have reduced the
longitudinal slope of the mainstream we call the Severn. Prior to the
Ironbridge diversion, the Stour–Avon drainage is recorded best by
the Avon terraces, while the pre-Wolstonian valleys have been so
deeply dissected by later erosion that only fragments remain, mostly
in those parts of the Welsh borderland valleys of the Lugg and Teme
that lay south of the Wolstonian ice. Near to the Severn itself the
highest, and thus presumably the oldest, gravels are called the Woolridge terrace (Wills 1938). Although earlier thought of as pre-Wolstonian in age, as the extent of the physical changes of the area has been

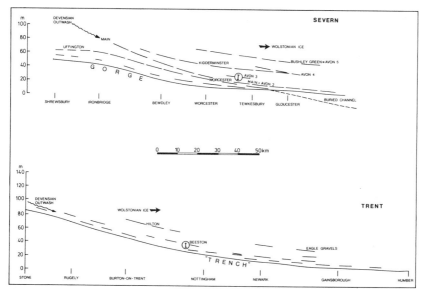

Fig. 12.3. Terraces of the River Severn and the River Trent. H = Hoxnian, I = Ipswichian.

recognized, it has seemed increasingly likely that the Woolridge gravels are no older than Wolstonian (Bishop 1957). They lie about 90 m above the modern floodplain, compared with 28–30 m for gravels of the same age in the Trent and Thames valleys, eloquent testimony to the major geomorphological changes that have affected the Severn basin, and an effective warning of the impossibility of deriving past sea-levels from the heights of river terraces above contemporary floodplains.

The Avon, although now a modest tributary of the Severn, should be described first. During Wolstonian times ice advanced south-westwards from the Leicester area and ponded a pro-glacial lake, Lake Harrison, (Shotton 1953) in the upper part of the former Soar valley. At first this was enclosed by the pre-Wolstonian water-parting from the Cotswolds to the Birmingham plateau, but later it was ponded higher against the Welsh ice occupying what is now the lower Severn valley. The Severn ice retreated before the Soar valley became ice-free, and a new river established itself by cutting back into the lake sediments, eventually shifting the divide 60 km north-eastwards to east of Bedworth. Vigorous incision into Lake Harrison sediments, the Lower Lias, and the Keuper Marl has allowed the Avon to develop a well-marked set of terraces, the rapid downcutting continuing with the general invigoration of the lower Severn after the Ironbridge diversion. The terraces have always been known as a numbered sequence from top to bottom, and No. 5 is contemporaneous with the first draining of the lake and the retreat of the Wolstonian ice.

The No. 3 terrace of the Avon lies on gravels which are generally believed (Shotton 1953) to be the early stage of an aggradation which culminated in the terrace gravels and flat of No. 4. The gravels at the level of No. 3 have a warm fauna, including *Hippopotamus*, and thus date from the Ipswichian interglacial. Most of No. 4 is much colder and marks the onset of the early phases of the Devensian glaciation. Shotton (1968) has noted evidence for the younger age of the gravels of No. 4 terrace at Brandon, which step across an underlying gravel regarded as Ipswichian. The No. 4 terrace lies at about 25 m above the floodplain, much higher than the equivalent terrace in the Trent. The No. 3 terrace is rather restricted in its occurrence, and this would be consistent with the suggestion that while the gravels at this level are primarily Ipswichian, the terrace flat must have been cut during the incision following the No. 4 stage, involving at least local reworking of the Ipswichian gravels. No. 2 terrace correlates with the Floodplain terrace of the Trent and the Main terrace of the Severn: as we shall see this marks the maximum advance of the Devensian ice into the Cheshire Plain. In the lower Avon, No. 2 terrace lies 10 m above the modern floodplain.

The Welsh Severn was obstructed by the Devensian ice and diverted by overflow (probably in part as an ice-marginal course) into the valley of the Stour. The diversion was well enough established to survive ice-retreat, and the Severn is now the trunk stream of the basin. The more steeply sloping Stour and Teme terraces prior to the diversion are the higher terraces; the Bushley Green which has been correlated with Avon No. 5, and the Kidderminster, correlated with Avon Nos. 3 and 4.

The Main terrace of the Severn is a remarkable feature, obviously formed by high discharges. Despite the relatively low slope the terrace gravel is extremely coarse (with many sizeable boulders), especially in the area immediately below Bridgnorth. The lithology, too, demonstrates the sudden accession of new erratic material from the Irish Sea and Welsh ice-sheet. Along the section from Bridgnorth to Worcester the Main terrace lies 30–33 m above the modern floodplain; by Tewkesbury the relative height has dropped to 16 m. Two lower terraces are recognized, the Worcester (which may be traceable through the Ironbridge gorge) and the Power Station terrace. The latter is really little more than an elevated part of the modern floodplain, but both illustrate the rule that if there is enough vertical separation of terraces in one river basin, it is likely that additional terraces will be recognizable, giving a fuller sequence than in a neighbouring basin where they are more closely spaced.

One of the most important Quaternary sites in the Severn basis is at Upton Warren. Here in the valley of the Salwarpe, a small left-bank tributary of the Severn (Coope, Shotton, and Strachan 1961), mammals, beetles, plant remains, and pollen point to an interstadial which was warm, but brief and treeless (see also Coope, Chapter 5).

Partly through the interest of the radiocarbon dating laboratory of the University of Birmingham, a number of dates have been obtained from the lower terraces of the Severn basin. While in general these confirm the correlations noted already, they pose some problems in interpretation, and have precipitated a good deal of discussion. Most of the dates on Avon No. 2 come out at about 31 000 BP (three dates from Brandon) and these correlate with a date of 32 160 BP from the low terrace of the Tame. However, material from Fladbury, lower down the Avon (60 km downstream from Brandon), is dated at 38 000 and the Upton Warren deposits have been dated at 41 500 and 41 900 in two separate determinations. Shotton (1968) interprets the 7000 year difference between Fladbury and Brandon as representing the upstream migration of aggradation: on the interpretation of river terraces put forward here it is a measure of the time taken for the knickpoint below No. 3 terrace to retreat to Brandon, an average distance of 9 m/year. The Upton Warren date is consistent with Fladbury and could represent some aggradation and connected ponding in response to the Main terrace of the Severn. The problem posed by this date has been its implication, via the Main terrace, for the date of the Devensian ice-advance to the Ironbridge gorge: evidence from the Cheshire Plain favours ice-retreat some time after 28 000 BP on the basis of the date from Four Ashes, north of Wolverhampton (Shotton 1967). Here again a river and its associated terraces, by providing a link between areas with relatively independent chronologies, provides a common datum with which both sequences must be reconciled.

Conclusion

As we have seen, rivers provide links between disparate areas, by flowing from one place to another and in time these links will provide a chain that will make increasingly secure our understanding of the chronology of Quaternary events. Some past correlations would have made rivers flow uphill: no doubt some equally absurd relationships remain to be exposed as we move towards a better understanding. But as a key to the fluctuation of Quaternary sea-levels, river terraces seem to have led us nowhere at all.

References

Bemrose, A.H.H. and Deeley, R.M. (1896). Discovery of mammalian remains in the old river gravels of the Derwent near Derby. *Q. Jl geol. Soc. Lond.* **52**, 497–510.

Bishop, W.W. (1957). The Pleistocene geology and geomorphology of three gaps in the Midland Jurassic escarpment. *Phil. Trans. R. Soc.* **241**, 255–306.

Briggs, D.J. and Gilbertson, D.D. (1973). The age of the Hanborough terrace of the River Evenlode, Oxfordshire, *Proc. Geol. Ass.* **84**, 155–74.

Clayton, K.M. (1953). The glacial chronology of part of the middle Trent basin. *Proc. Geol. Ass.* **64**, 198–207.

Coope, G.R., Shotton, F.W. and Strachan, I. (1961). A late Pleistocene flora and fauna from Upton Warren, Warwickshire, *Phil. Trans. R. Soc.* **244**, 379–421.

Franks, J.W. (1960). Interglacial deposits at Trafalgar Square, London, *New Phytol.* **59**, 145–52.

Hare, F.K. (1947). The geomorphology of part of the middle Thames, *Proc. Geol. Ass.* **58** (4), 294–339.

Jones, P.F. and Stanley, M.F. (1974). Ipswichian mammalian fauna from the Beeston terrace at Boulton Moor, near Derby, *Geol. Mag.* **111** (6), 515–20.

King, W.B.R. and Oakley, K.P. (1936). The Pleistocene succession in the lower part of the Thames valley, *Proc. prehist. Soc.* **2** (1), 52–76.

Leopold, L.B., Wolman, M.G. and Miller, J.P. (1964). *Fluvial processes in geomorphology*. W.H. Freeman and Company, San Francisco and London.

Miller, A.A. (1939). Attainable standards of accuracy in the determination of preglacial sea levels by physiographic methods, *J. Geomorph.* **2**, 95–115.

Posnansky, M. (1960). The Pleistocene succession in the Middle Trent Basin, *Proc. Geol. Ass.* **71** (3), 285–311.

Schumm, S.A. (1965). Quaternary paleohydrology. In *The Quaternary of the United States*, (ed. H.E. Wright and David C. Frey), pp. 783–94. Princeton University Press, New Jersey.

Shotton, F.W. (1953). The Pleistocene deposits of the area between Coventry, Rugby and Leamington and their bearing upon the topographical development of the Midlands, *Phil. Trans. R. Soc.* **237**, 209–60.

Shotton, F.W. (1967). Age of the Irish Sea glaciation of the Midlands, *Nature, Lond.* **215**, 136.

Shotton, F.W. (1968). The Pleistocene succession around Brandon, Warwickshire, *Phil. Trans. R. Soc.* **254**, 387–400.

Swinnerton, H.H. (1937). The problems of the Lincoln gap, *Trans. Lincs. Nat. Un.* **9**, 145–53.

Waechter, J. d'A., Newcomer, M.H. and Conway, B.W. (1970). Swanscombe (Kent–Barnfield Pit) 1971. *Proc. R. anthropol. Inst. for 1970*, 43–64.

West, R.G. (1969). Pollen analysis from interglacial deposits at Aveley and Grays, Essex, *Proc. Geol. Ass.* **80**, 271–82.

West, R.G., Lambert, J.M. and Sparks, B.W. (1964). Interglacial deposits at Ilford, Essex. *Phil. Trans. R. Soc.* **247**, 185–212.

Whitaker, W. (1889). The geology of London, Vol. 1. *Mem. geol. Surv. U.K.*

Wills, L.J. (1924). The development of the Severn valley in the neighbourhood of Ironbridge and Bridgnorth, *Q. Jl geol. Soc. Lond.* **80**, 274–314.

Wills, L.J. (1938). The Pleistocene development of the Severn from Bridgnorth to the sea, *Q. Jl. geol. Soc. Lond.* **94**, 161–242.

Wooldridge, S.W. (1938). The glaciation of the London Basin and the evolution of the lower Thames drainage system. *Q. Jl geol. Soc. Lond.* **94**, 627–67.

Wooldridge, S.W. (1958). Some aspects of the physiography of the Thames in relation to the Ice Age and Early Man, *Proc. prehist. Soc.* 23, 1–19.

Wooldridge, S.W. and Linton, D.I. (1955). *Structure, surface and drainage in south-east England.* George Philip, London.

Zeuner, F.E. (1945). The Pleistocene period: its climate, chronology and faunal successions, (1st ed.) *Ray Soc. Monograph* 130.

Zeuner, F.E. (1959). *The Pleistocene period: its climate chronology and faunal successions,* (2nd ed.) Hutchinson, London.

Zeuner, F.E. (1961). The sequence of terraces of the lower Thames and the radiation chronology. *Ann. N.Y. Acad. Sci.* 95, 377–80.

13

Raised beaches and sea-levels

G. F. MITCHELL

Abstract

After recognizing the rapidly growing importance of the study of deep-sea cores and of global hydro-isostatic deformation, this chapter gives a conventional review of raised beaches and sea-levels around the British Isles from the late Pliocene to today. When considering the return of water to the oceans after 15 000 BP, as the last ice-masses melted, a relatively steady rise in level is envisaged, as shown in curves of the Emery type, rather than a fluctuating rise, as shown in curves of the Fairbridge type. The controversial claim is made that about 5000 BP the sea around the British Isles stood at about 4 m above its present level.

Introduction

While it seems appropriate to give a conventional and largely parochial review of what appears to be known at present of sea-levels around the British Isles during the Quaternary, I am well aware that our insular problems are only a very small part of what is essentially a global problem and one which is already being attacked on a world-wide scale.

And as we look at our local problems we must have in our ears the warnings recently reiterated by Thom and Chappell (1976). The relative movements in sea-level that took place in Quaternary time resulted from a number of factors which operated either on a relatively local basis (regional flexures and/or faulting, hydro-isostatic deformation of the continental shelf and the adjacent coasts, meteorological changes affecting the magnitude and frequency of storminess, changes in circulation and tidal amplitude induced by sedimentation within partially enclosed water bodies), or on a wider scale (glacio-eustatic, and global hydro-isostatic deformation).

Recent mathematical studies of global deformation (Farrell and Clark in press) suggest that even when global water volume was increasing consequent on ice-melting, some areas of the globe might experience lower water levels.

To complicate interpretation further, in and around the British Isles we have at least three provinces (Fig. 13.1), each of which behaved in a different way. First we have the North Sea basin where marked tectonic sinking—which still continues—throughout the Quaternary has allowed thousands of metres of deposit to accumulate. Second we have the stable Celtic Sea area, along whose drowned coastline we can trace a wave-cut shore-platform, whose level is remarkably horizontal, and which must be of considerable antiquity. Third is the northern part of Great Britain and Ireland, where repeated formation of great ice masses brought about isostatic sinking of the land on a scale that has hitherto been seriously underestimated; some areas may be still rising in continuing recovery.

The deposits that are available for study may thus have histories that are both very complicated and very different, and it would be naive indeed to assume that beaches which are at the same level must necessarily be of the same age.

The Cainozoic background

The Celtic Sea is bordered by surviving blocks of the Hercynian massif in southern Ireland, south-west Wales, Cornwall, and Brittany and these may have been essentially stable throughout the Cainozoic. Their basic rock topography and the course of the river valleys had

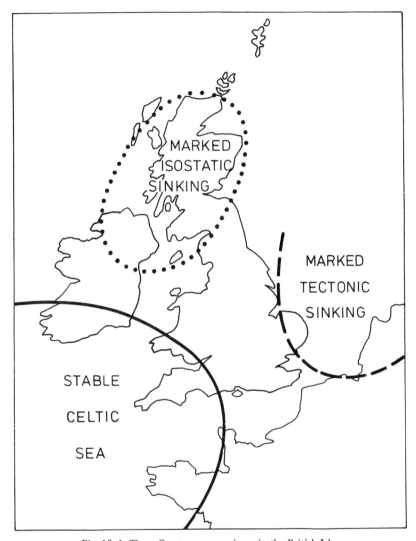

Fig. 13. 1. Three Quaternary provinces in the British Isles.

already attained much of their present configuration before the Pliocene came to an end. Today their coasts are clearly 'drowned', and at the time of the Pliocene the drowning may well have been still more severe.

In Cornwall the marine clay of late Pliocene age at St. Erth reaches a height of + 35 m, and probably indicates a sea level of + 45 m (Mitchell *et al.* 1973). In Brittany the marine clays in the Guindy valley, which indicate a sea-level of about + 60 m (Morzadec-Kerfourn 1975) are of about the same age, or perhaps a little older.

In the North Sea basin also sea-level in late Pliocene time was above its present height. Much of the Netherlands was covered (Zagwijn 1974) and the sea also lapped over into East Anglia where Coralline Crag was deposited.

We have no knowledge of events in the northern part of the British Isles.

The Quaternary before 700 000 BP (pre-Brunhes Epoch)

In a deep boring in the central Netherlands the Pliocene–Pleistocene boundary lies in a marine deposit at a depth of about 300 m and has an age of about 2·5 million years (Zagwijn 1975). The boundary is defined by the first appearance of the cold water Foraminifer, *Elphidium oregonense*, at present an inhabitant of the Bering Straits. In the south-eastern part of the Netherlands, the Pliocene and lowermost Pleistocene beds appear in a non-marine facies.

In East Anglia the boundary is placed at the base of the Waltonian stage of the shallow-water marine Red Crag (Mitchell *et al.* 1973), where the Red Crag rests almost certainly unconformably on the Boytonian facies of the Pliocene Coralline Crag, also a shallow-water marine deposit.

All round the Irish Sea, tills and outwash gravels incorporate Red Crag shells, particularly richly at Killincarrig, Co. Wicklow (McMillan 1938) where the fauna resembles that of the Little Oakley horizon. The Irish Sea must have been marine at the opening of the Pleistocene.

For the long stretch of time, perhaps 1·8 million years, that follows until the opening of the Brunhes Epoch, we have scanty information. Tectonic movements probably continued in the Netherlands throughout the period. In eastern England there was post-Red Crag movement, because the crag ranges from +200 m to −50 m. Marine conditions continued in East Anglia, but there is evidence of Lower Pleistocene marine regression in the Netherlands.

The temperature curve from the Netherlands (Zagwijn 1975) now spans the whole of the Pleistocene, and shows long periods with low temperatures throughout. The cold Eburonian stage about 1·5 million years ago has been equated with the Nebraskan ice-mass in North America. It was presumably a time of low sea-level but whether ice-masses expanded in all the other stages, or whether some of these stages saw periglacial activity only, we cannot tell.

The Quaternary after 700 000 BP (Brunhes Epoch)

The pre-Hoxnian Stages

Zagwijn (1975), in his classification of the Netherlands, has a long 'Cromerian Complex' starting just before the Brunhes Epoch and lasting until about 350 000 years BP. Shackleton and Opdyke (1973), working on deep ocean cores, have extended their foraminiferal oxygen isotope record to below the base of the Brunhes and interpolated a time-scale based on a constant rate of foraminiferal sedimentation. On their premise 'that ice volume, or sea-level, may be reliably estimated from the isotope data' and the further assumption that ice accumulation means a cold climate, several alternations between heat and cold can be inferred for the 'Cromerian Complex', with attendant changes of sea-level. Zagwijn's pollen diagrams also show a succession of climate changes, though there is no close correspondence with the isotope curves. However, Zagwijn also believes that no sedimentation during the earlier three quarters of the Dutch 'Cromerian Complex' has been recognized in East Anglia and he places the Pastonian warm stage, the Beestonian cold stage, and the Cromerian (s.s.) warm stage of West (1968) at the end of his complex. West records tidal silts in both the Pastonian and the Cromerian warm stages, and this suggests a sea-level not too different from that of today. Indeed from this point on we get a clear impression that in each warm stage sea-level returned to approximately the same level. The wave-cut shore-platform that surrounds the Celtic Sea at a height of about + 4 m with the cliff notch rising in places to + 10 m may have been fashioned during this relatively long period.

Our first clear record of ice-masses in Britain comes in the Anglian Cold Stage, which may correspond with the Elsterian, and immediately precedes the Hoxnian Warm Stage. There are two till complexes in East Anglia, a lower Cromer Till, and an upper Lowestoft Till, and these are separated by the Corton shelly sands and gravels. The status of the Corton Beds has long been in doubt, but they do contain seams of arctic freshwater clay, with occasional brackish Foraminifera. We can perhaps picture pro-glacial braided outwash gravels merging into a delta in an arctic sea at about + 20 m (West and Wilson 1968).

The Hoxnian Warm Stage

By the time of the opening of the Hoxnian Warm Stage, perhaps about 250 000 years ago, one could hope that firmer ground is being reached. The apparently complete pollen record from Marks Tey (Turner 1970) provides a reference standard, and the less full record from Hoxne itself gives archaeological cross-references.

With regard to sea-level, the estuarine silts at + 27 m at Kirmington in Lincolnshire (Watts 1959) have a pollen content that suggests a Hoxnian age, and implements in the overlying gravel support this. Marine clay at + 20 m in the Nar Valley, Norfolk (Stevens 1960) again has a pollen content suggesting a Hoxnian age. In the lower Thames valley at Swanscombe (Waechter 1971) terrace deposits between + 23 and + 30 m have an archaeological content consistent with the Hoxnian Warm Stage. If we set aside the possibility that there may have been later subsidence along the western margin of the North Sea basin, then these sites appear to point to a Hoxnian sea-level of the order of + 25 m.

But Shackleton and Opdyke point out that if we accept their interpretation of the oxygen-isotope record as an ice-volume or ocean-volume record, then the Hoxnian ocean-volume was less than that of the succeeding warm stage, the Ipswichian or Eemian. As we shall see later, Eemian sea-level seems to have reached about + 8 m, and therefore Shackleton and Opdyke claim that if there was an older warm stage sea-level of + 25 m, it cannot be of Hoxnian age, but must be older.

Attack also comes from another direction. The gravel ridge which occurs at + 30 m at Slindon, Sussex, and at other points along the south coast of England, has generally been taken to be a storm beach ridge of Hoxnian age, like the modern high beach ridge at Portland Bill. Its structure, its temperate fauna, and its implements suggest it to be the equivalent of the Swanscombe gravels.

But it has recently been suggested (Kellaway *et al.* 1975) that the gravels are of fluvioglacial origin, and derive from ice in the English Channel. Stones from distant sources (erratics) have long been known from the shores of the English Channel, and ice-rafting has been thought to be the method of the emplacement. Kellaway and his colleagues on the other hand consider that on at least two occasions great masses of ice formed in the basin of the Celtic Sea, and from there flowed north-eastwards up the English Channel. If ice from such a source did reach the Straits of Dover, it had travelled upslope for a distance of 600 km. As the parent ice-mass had a much steeper downslope freely accessible on its seaward side, it must have been of truly enormous thickness if it was able to set up a gradient of flow which could carry ice as far as the English Channel.

Kellaway and his colleagues also minimize the evidence for eustatic beaches at *c.* + 25 m round the shores of the Celtic Sea, and hint at isostatic effects produced by their postulated ice-sheets. But along the south coast of Brittany the 25 m level is consistently reached by the unfortunately named 'Haut-Normannien' beach, and there is a storm beach at 40 m at St. Helier in Jersey. In south-west England a + 20 m beach is described at Penlee, outside Penzance, and at Hughtown, St. Mary's, Isles of Scilly, there is a storm beach at + 25 m.

In all these areas the shore-platform is horizontal, and if it is pre-Hoxnian in age, it shows no signs of isostatic deformation. In defiance therefore of Shackleton and Opdyke, and Kellaway *et al.*, this chapter continues to consider that in the Hoxnian Warm Stage sea-level did reach + 25 m.

When the peak of the eustatic transgression had passed, and the growth of the incipient Wolstonian ice-masses began to pull sea-level down, the surface waters got colder, and pack-ice with embedded erratics could float farther and farther south. What was the level of the sea when floating pack-ice could first circulate in the English Channel?

It has long been noted that the fossil beach at + 12 m at Portland has a relatively cool fauna and an unusually high proportion of erratics among its gravels (Arkell 1951); can it be that by the time the sea had dropped to this level, it was already cold enough to transport ice-floes? We are on stronger ground in south-west Ireland. The shore-platform is very well developed on the north shore of Tralee Bay, and as sea-level at the end of the Hoxnian Warm Stage fell below its present level, beach ridges were abandoned on the platform. Peats and silts formed between the ridges, and their pollen content showed that by the time they were forming, the Hoxnian woodlands had disappeared and Wolstonian solifluction had begun (Mitchell 1970). At Kilmorequay in Wexford the beach gravels on the platform were cryoturbated before Wolstonian head crept down over them.

This chapter thus takes the view that in the late Hoxnian it was possible for ice-floes to emplace erratics on the older wave-cut shore-platform, before sea-level had dropped so far as to make the platform inaccessible to floating ice. The case for an immense glacier flowing from west to east up the English Channel is regarded as not proven.

We should end on a note of caution. As sea-level fell from its maximum of + 25 m, it was possible for beach pebbles to be emplaced wherever rock topography provided a lodging-place, or an embayment gave opportunity for a spit to grow. The notch between the old shore-platform and the cliff behind provided a particularly favourable resting-spot, and beach pebbles accumulated here at many localities. We thus have a sequence of Hoxnian 'beaches', straggling down the shore-line at different elevations, and we have to agree with Monnier (1974) that if we know nothing more about a fossil littoral deposit than its altitude, we cannot date it.

The North Sea existed in Hoxnian time, because late Hoxnian marine sediments were found at Lat. 53°N. and Long. 1°E. at a depth of −50 to −55 m. The fossils suggested a water depth of about 20 m; if the sediments were *in situ*, then post-Hoxnian subsidence would appear to be indicated (Fisher *et al.* 1969).

The Wolstonian Cold Stage

The oxygen-isotope curve suggests that large masses of ice must have formed between about 200 000 BP and 130 000 BP, and the wide distribution of glacial deposits of Wolstonian age in north-west Europe would seem to confirm this suggestion. The Netherlands were invaded by ice; Great Britain north of a line from Bristol possibly to London (but others might claim only to North Norfolk) was probably completely covered; almost the whole of Ireland seems to have been covered at one stage or another. But while glacial deposits are well known, there are no beaches and only scanty evidence of marine deposits of Wolstonian age.

An important feature of the Wolstonian Cold Stage was the advance of an enormous *mer-de-glace* down the basin of the Irish Sea. Much of the ice must have formed north of Glasgow, because as it advanced south it overrode Ailsa Craig in the Forth of Clyde, and assimilated large quantities of the riebeckite-bearing microgranite of which the island is formed. Erratics of this very characteristic rock were carried as far as south Ireland and south Wales. Ice swept along the south coast of Ireland as far as Cork harbour, and removed most of the late Hoxnian beach deposits. Some patches did survive where rock topography gave protection from the eroding ice.

Even before this advance of ice from Scotland, ice had formed in central Ireland, and from there had advanced eastwards into the basin of the Irish Sea near Drogheda (Colhoun and McCabe 1973). This ice-mass must have initiated isotatic subsidence. As the much greater ice-mass built up over Scotland subsidence must have accelerated, and even though sea-level was falling glacio-eustatically, the sinking of the land enabled sea-water, now of the nature of an arctic fjord, to remain in the Irish Sea. An arctic marine fauna, indicating a water-depth of at least 100 m, was present in the fjord, and floating ice dropped erratics into the fine-grained sediments on its floor. Ultimately the Scottish ice on its way southwards obliterated the fjord, and deposited a till rich in marine clay and shelly debris on top of the fjord deposits. After the ice had melted away, isostatic uplift raised the fjord deposits to a height of + 35 m, the height at which they occur today. A total vertical movement of at least 135 m is indicated. Some critical voices have suggested that the fjord material is an erratic in the Scottish till, but the field evidence gave no support to this view.

Round the shores of the Celtic Sea beyond the limits of the Scottish ice, the beaches that had been stranded as Hoxnian sea-level fell were buried below periglacial deposits, either head or loess. In the southern part of the Scilly archipelago the late-Hoxnian Chad Girth beach (which contains only very rare erratics of flint, likely to have been derived from the chalk of the surrounding sea floor and brought in from a distance by ice) is buried by Lower Head which contains only very rare erratics (Mitchell and Orme 1967).

At the famous section at Valais à Cesson, near Saint-Brieuc in Brittany, the late-Hoxnian beach is overlain by sandy loam with seams of pebbles; some of the silt-size particles in this solifluction-deposit are probably of loessic origin (Monnier 1974).

The Ipswichian Warm Stage

The oxygen-isotope curve suggests that there was a considerable melting of the world's ice in the Ipswichian Warm Stage, and that the highest sea-levels came early in the stage. Sparks and West (1972) give a considerable discussion of Ipswichian events in the south of England; they envisage that maximum sea-level (about + 8 m over today's level) came relatively late in the stage, i.e. after the wood-lands had passed their climax phase. They discuss the possibility of base-level oscillation, and show that there was downwarping towards the North Sea Basin. In the Netherlands the contemporary marine Eemian deposits lie below a cover of 10–20 m of younger deposits.

The famous beach at Black Rock, Brighton (where the beach shingle reaches to + 12 m), and its counterpart at Sangatte, Pas-de-Calais, are generally accepted as of Ipswichian age, as are the fossili-ferous mud and overlying beach at Selsey, Sussex. But as we move west towards the Celtic Sea, controversy rises.

At Valais à Cesson (already referred to) the loams overlying the Hoxnian beach contain a palaeosol regarded as Ipswichian in age, but at other sites in Brittany the Ipswichian waves cut a platform in the Wolstonian loams and deposited a beach on the platform at a height of about + 5 m. Such a beach is well seen at Ruisseau de Vaux, Plèvenon, Côtes du Nord, and at Portelet Bay in Jersey.

At Porth Seal, St. Martin's, Isles of Scilly, a similar platform trims not only Wolstonian head (itself resting on beach materials of Hox-nian age), but also the old shore-platform. As the Wolstonian ice had brought much erratic material to the vicinity of the Scilly Isles, the Ipswichian beach contains much foreign material, in marked contrast to the lower Hoxnian beach.

I believe that the Gower Peninsula in Wales also holds two beaches, an older Hoxnian beach poor in erratics, and a younger Ipswichian beach rich in derived Wolstonian erratic material. Unfortunately the two beaches are at about the same level, and are not seen anywhere superimposed, and separated by head, as in the Scilly Isles. This interpretation is strenuously contested by Bowen (1973), who claims that there is only one beach, of Ipswichian age, in Gower. He also claims that the beach of the south Irish coast is also of Ipswichian age, and that the overlying till is of Devensian age.

Dispute also arises in Barnstaple Bay, where a beach deposit resting on the shore-platform is covered by glacial material. Stephens (1970) sees this as Wolstonian till resting in primary position on a Hoxnian beach; Kidson and Wood (1974) see it as Wolstonian till soliflucted on to a younger Ipswichian beach by Devensian freeze-thaw processes.

As we move up the Irish Sea, the Ipswichian trail peters out. A thin organic deposit at West Angle in Milford Haven (John 1970) may be of Ipswichian age. At Poppit, Cardiganshire, and again at Cahore, Wexford, beach deposits lie below Devensian Till; the beaches may be of Ipswichian age.

In the tidal part of the Severn Estuary the shallow water shelly Burtle Beds rise to a height of +12 m (Kidson and Heyworth 1976).

Along the North Sea coast, the Easington beach gravel in County Durham has been suggested as Ipswichian, but its height of *c.* + 18 m raises doubts. Further south at Sewerby, an undoubted Ipswichian beach and cliff notch, with an important fauna including *Hippopotamus* and *E. antiquus*, is only at + 2 m. Still further south at Tattershall in south Lincolnshire, an inland late Ipswichian deposit with land snails and plants stands only about a metre above present sea-level. It is probable that post-Ipswichian subsidence of the south North Sea basin has taken place.

The Devensian Cold Stage

At a date perhaps a little younger than 80 000 years ago the oxygen-isotope curve moves steeply, and suggests the rapid build-up of ice in the northern hemisphere (Shackleton and Opdyke 1973, Fig. 7). Ice certainly formed in the St. Lawrence lowland, because the organic deposits of the St. Pierre Interstade, sandwiched between two tills, have been suggested by carbon-14 dating to have an age of about 65 000 years (Dreimanis and Goldthwait 1973). The controversial question at once arises 'Did ice form in Europe—in the British Isles—at this time?'. In Britain, no Devensian till has been recognized which is not Late Devensian. In Ireland the Middle Devensian deposit at

Derryvree (Colhoun *et al.* 1972) rests on a till which does not appear to have been substantially weathered or cryoturbated after its deposition, and this might be of Early Devensian age.

Sea-level will have fallen substantially in this early cold period. The oxygen-isotope record then suggests that it recovered somewhat in level about 50 000 years ago, and such a move would be in harmony with the milder climates of the Middle Devensian Upton Warren interstadial complex.

The big Late Devensian build-up of ice then followed, and with radiocarbon dates on firmer ground, we can trace the movement of sea-level in some detail. The lowest level, about -130 m, was attained about 15 000 years ago. Although the ice-masses of Devensian age in the British Isles were substantially smaller than the Wolstonian masses, they tended to follow the same pattern, which was to build up over Scotland, and then to move down the North Sea basin, and down the basin of the Irish Sea. The lowering of sea-level meant that in most of these areas the former sea floor was dry, and the ice-masses were grounded rather than floating ice-shelves. When the ice started to decay, linear hollows, perhaps of the nature of tunnel-valleys, were cut into the underlying floor.

The Late Devensian (26 000 to 10 000 BP)

Stratigraphical terminology. It is desirable to label at least 3 stages of the Late Devensian (Late Midlandian of Ireland). The earliest stage when great masses of ice flowed and ebbed might in Ireland be called the Glastry Stadial, after a site in County Down (Mitchell *et al.* 1973, p. 73). Here there are two successive drumlin-forming tills, the lower containing shells dated by carbon-14 at 24 000 BP. The upper till at least must be late Devensian, and with later work it may be possible to sub-divide the Glastry Stadial.

Evidence from many sources now points to a phase of climatic amelioration of interstadial rank covering the upper part of Pollen Zone I and the whole of Zone II. This stage has been named the Woodgrange Interstadial in Ireland (Mitchell 1976), and the Windermere Interstadial in England (Pennington 1975, Coope and Pennington 1977). It might be called Clyde in Scotland (Bishop and Dickson 1977).

The Irish Nahanagan Stadial (Mitchell 1976) and the Loch Lomond Stadial of Scotland equate with Pollen Zone III.

The Glastry Stadial (26 000 to 14 000 BP).

If we judge by the amount of isostatic depression produced, the Late Devensian ice-mass—like its Wolstonian predecessor—must have been thickest in central Scotland, where the down-sinking was in excess of 250 m. We cannot as yet determine the outer limits of the area that did undergo some depression, but if we connect all points where the earliest late-glacial beach is at present sea-level, we encircle an area where isostatic uplift has exceeded the postglacial glacio-eustatic rise in sea-level (see Fig. 13.2).

At this point I must make clear my position with regard to the various curves that have been published in recent years, purporting to

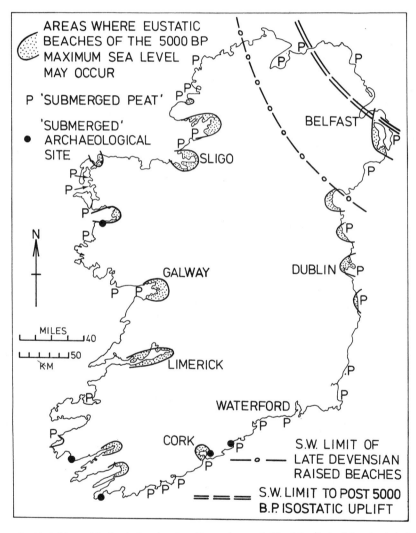

Fig. 13.2. Map of Ireland showing late Devensian and Flandrian 'raised beaches', 'submerged peats', and 'submerged' archaeological sites.

trace the rise in sea-level that was consequent on the melting of the Late Devensian ice-masses. I believe that after 15 000 BP there was an essentially continuous rise in sea-level, and that this rise continued till the sea reached a height of $+4$ m at about 5000 BP; it later fell below present level, and subsequently rose again to it (Fig. 13.3). In other words I tend to follow the Emery curve as modified by Labeyrie (see Mitchell and Stephens 1974, Fig. 1), and to reject the oscillating curves of Fairbridge, Mörner, Tooley, and others.

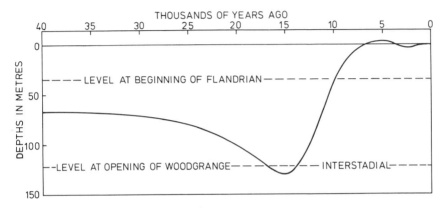

Fig. 13.3. Curve to suggest the movement of sea-level from the middle Devensian to today.

The sea-level curve makes it clear that after 15 000 BP the ice-masses were on the retreat, and the rising sea began to flood in over lands that were still isostatically depressed. The peripheral parts of Scotland were extensively invaded as the sea pressed up to the margins of the retreating ice, and gave rise to complex deposits where braided outwash terraces merged into extensive intertidal deltaic flats. Beaches now $+30$ m above sea-level may have formed when world-wide sea-level was below -100 m, indicating uplift of at least 130 m.

The Woodgrange Interstadial (14 000 to 10 600 BP). In the estuary of the Clyde the occurrence of deposits with a relatively warm late-glacial fauna sandwiched between colder deposits in a marine sequence has long been known. These deposits seem to offer an important opportunity to define this Late Devensian interstadial in a marine environment. In the Paisley area shells with an age of about 13 000 years occur in clays which rise to $+25$ m (Bishop and Dickson 1970); further downriver near Dumbarton, material about 1000 years younger lies at about -30 m (Peacock 1971); early isostatic uplift must have been very rapid. Beds with similar fauna also occur on the opposite side of Scotland in the Cromarty Firth, at a depth of about −50 m (Peacock 1974).

Across the North Channel in Ireland, at Roddans Port, Down, freshwater interstadial muds with an age of 12 000 years lie today at modern sea-level, and the same situation has recently been reported from further north, at Carnlough, Antrim (Prior and Holland 1975). Presumably rapid recovery from a smaller degree of isostatic depression than that in Scotland enabled these deposits to escape submergence by the late-glacial sea. Nonetheless the downward movement of the north of Ireland was not inconsiderable, as late-glacial beaches occur at Malin Head, Donegal, at an altitude of +20 m.

During the interstadial there were extensive land-bridges between Britain, the Isle of Man, and Ireland (Mitchell 1976, p. 69), and extensive movements of plants and animals took place.

The Nahanagan Stadial (*10 500 to 10 000 BP*)

While in Ireland only minimal amounts of ice seem to have formed in the Nahanagan Stadial (Colhoun and Synge, in press), in Scotland there was quite an extensive ice-cap, which extended as far south as Loch Lomond (Sissons 1974). Some glaciers flowing away from the ice-cap ended in the sea, while others ploughed up the earlier interstadial deposits. It may be that the new ice-cap was of sufficient weight to retard isostatic recovery, and the rising sea may have transgressed still farther into Scotland.

There are a great many sites available for study, and when these have been examined in detail, the events of the stadial should become reasonably clear. Periglacial conditions in the stadial were sufficiently severe for cryoturbation structures to develop, and these are of help in sorting out the history of beaches associated with isostatic recovery. Beach deposits older than the stadial may show cryoturbation features; younger deposits will not.

The Flandrian Warm Stage before 5000 BP

Except in Scotland and the most extreme north-east corner of Ireland, isostatic recovery must have been largely complete by 10 000 BP, when the Flandrian Warm Stage opens. Sea-level was still at least 35 m below its present level, but was continuing to rise. Some landbridges still existed.

The rise from −35 m at 10 000 BP to +4 m at 5000 BP is attested by numerous radiocarbon dates. There is 8350 BP from Boreal peat with *Corylus* and *Pinus* at −32 m in the English Channel off Cap Gris-nez. In Morecambe Bay in the east Irish Sea, sea-level rose from

−20 to −15 m between 9000 and 8000 BP (Tooley 1974). Peats and woods between present tide-marks have the following ages; in Lough Foyle, Donegal, a *Quercus*-trunk was 6955 BP; a Boreal peat near Bray, Wicklow was 6760 BP; a Boreal peat at Westward Ho in Devon was 6585 BP.

Evidence from Sutton, Dublin (Mitchell 1972) indicates that sea-level reached its Flandrian maximum of +4 m about 5000 BP. Howth Head, on the north shore of Dublin Bay was then an island, and mesolithic people were still living on its sheltered western shore at Sutton, where their shell-debris built up a midden; charcoal from the midden had a carbon-14 age of 5250 BP. At the maximum of the transgression (+4 m) a storm washed away part of the midden, but the Mesolithic people continued to occupy the site, and threw debris on the storm beach, which was abandoned by the falling sea. In the inner reaches of Strangford Lough, Down, where there cannot have been any significant wave action, marine deposits rise to +3·5 m (Mitchell and Stephens 1974).

The Midland Valley of Scotland in the vicinity of Stirling was long occupied by a marine fjord, but continuing, though slowing, uplift gradually raised the area above sea-level, and a layer of freshwater peat formed; the base of the peat had an age of 8700 years. The rate of sea-level rise then overtook the rate of isostatic uplift, and the sea flooded in over the peat margins, and laid down a clay, generally known as carse clay; the top of the peat underlying the clay had an age of 8300 years. As long as the sea continued its eustatic rise, clay continued to accumulate; but as soon as the sea started to drop in level, freshwater peat developed on its surface; the base of the peat had an age of 5500 years. At this time sea-level was at +4 m; the highest level of the carse clay is at +15 m, and therefore this part of Scotland must have risen isostatically by about 10 m in the past 5000 years.

The concept that sea-level continued its eustatic rise to a height of 4 m above its present level is contested by Shackleton (personal communication), who claims that there is no isotopic evidence for massive ice-melting at this time. But in my opinion Flandrian ice-melting continued through the 'climatic optimum' (hypsithermal interval), and did raise sea-level to this height. The later decline in temperature towards the 'Little Ice Age' allowed the remnant ice-masses to start to grow again, and to draw sea-level back from the climax-level it had attained.

In the past beaches associated with the eustatic rise to +4 m have been confused with late isostatically raised beaches. In Ireland an effort has been made to disentangle the two, and this is illustrated in Fig. 13.2.

In the North Sea area there is in the Netherlands a puzzling series of oscillations, which first bring in marine clay over peat, and then allow a later peat to colonize the surface of the clay. The older series, the Calais Series, ranges from 7000 to 4000 BP (Zagwijn and van Staalduinen 1975). It is not clear if the transgressions and regressions reflect rises and falls in sea-level.

In north-west England (Tooley 1974) similar transgressions lie between 9270 and 4800 BP, and these are interpreted as due to oscillating sea-level.

The Flandrian Warm Stage, 5000 BP to today

At Sutton, Dublin, as already noted, a maximum Flandrian level at +4 m is dated to 5250 BP. The falling waves linked the former island of Howth to the mainland by a tombolo, and shells from this gave dates of 4830 and 4460 BP. Storm beaches were then banked against the flanks of the tombolo, and peat from a depression between two beach ridges was dated at 3730 BP. There is a group of dates from peats and woods at intertidal level today on the south and west coasts of Ireland that centres on 4000 BP; St. Margaret's, Wexford, 4185 BP; Ballycotton, Cork, 4100 BP; Fethard, Wexford, 4025 BP; Spiddal, Galway, 3750 BP.

There are also a number of archaeological sites between tide-marks (see Fig. 13.2); they range in age from a dolmen at Rostellan, Cork (possibly late Neolithic, possibly an antiquarian folly), to a crannog at Ardmore, Waterford (perhaps eighth century A.D. in age). Thus in Ireland there seems to be evidence that after 5000 BP the sea fell back to below its present level, and then recovered again. Such movement may have taken place more than once, but how many movements there were, what levels they reached, and when they took place is far from clear.

The same confusion reigns elsewhere. On the south side of the Celtic Sea in north Brittany (Morzadec-Kerfourn 1974) a series of dates from intertidal peats and woods ranges from 4250 to 2180 BP. Again between tide-marks, there are archaeological sites ranging from megaliths (5000 to 4000 BP) to salt-pans (2000 BP). Partly submerged cist-graves are known in the Isles of Scilly.

In the Netherlands the record of oscillations continues as the Duinkerke Series, 3500 to 1200 BP. In East Anglia a marine incursion deposited fen clay between 4600 and 3400 BP; as peat below the fen clay is now at −6 m, downwarping must be involved. Peat overlying the fen clay, and exposed on the modern foreshore, was dated to 2350 BP; there is also evidence of marine transgression in Romano-British times. At Maldon in Essex, salt-pans have been dated to 2130 BP.

In western England, there appears to have been a marine transgression in the Somerset Levels in Romano-British time. In the northwest the oscillations in level continued from 3770 to 1380 BP.

From such records as I have seen, it would appear that young 'submerged peats' have not been identified in the area in which uplifted late-glacial beaches occur. It would appear that in the latter area continuing isostatic recovery has been sufficient to hold the area above the late fluctuations in sea-level that appear to have occurred.

As I said at the opening of this chapter, studies of sea-level, of beaches, and of coral reefs are currently undergoing vigorous and exciting growth, and I am glad to acknowledge stimulating discussion with Professor Art Bloom of Cornell and Professor Roland Paepe of Brussels. Had this review been prepared for the XIth INQUA Congress, and not for the Xth in Birmingham in 1977, it might have had a very different tale to tell.

References

Special Report No. 4 of the Geological Society of London, 1973, *A correlation of Quaternary deposits in the British Isles*, by G.F. Mitchell, L.F. Penny, F.W. Shotton, and R.G. West, contains a very full bibliography. Publications cited in that bibliography, and which are referred to in this paper, are not included in this bibliography.

Arkell, W.J. (1951). Dorset Geology, 1940–1950. *Proc. Dors. nat. Hist. archaeol. Soc.* 72, 176–94.

Bowen, D.Q. (1973). The Pleistocene Succession of the Irish Sea. *Proc. Geol. Ass.* 84, 249–72.

Colhoun, E.A. and McCabe, M. (1973). Pleistocene glacial, glaciomarine and associated deposits of Mell and Tullyallen townlands, near Drogheda, eastern Ireland. *Proc. R. Ir. Acad.* B 73, 165–206.

Colhoun, E.A. and Synge, F.M. (in press) Lateglacial deposits at Lough Nahanagan, Co. Wicklow.

Coope, G.R. and Pennington, W. (1977). The Windermere Interstadial of the Late Devensian. *Phil. Trans. R. Soc. Lond.* B 208.

Dreimanis, A. and Goldthwait, R.P. (1973). Wisconsin Glaciation in the Huron, Erie, and Ontario Lobes *Mem. Geol. Soc. Am.* 136, 71–105.

Farrell, W.E. and Clark, J.A. (in press) On postglacial sea-level.

Fisher, M.J., Funnell, B.M. and West, R.G. (1969). Foraminifera and Pollen from a Marine Interglacial Deposit in the Western North Sea. *Proc. Yorks. geol. Soc.* 37, 311–20.

Houbolt, J.J.H.C. (1968). Recent sediments in the southern bight of the North Sea. *Geologie. Mijnb.* 47, 245–73.

Kellaway, G.L., Redding, J.H., Shephard-Thorn, E.R. and Destombes, J.-P. (1975). The Quaternary history of the English Channel. *Phil. Trans. R. Soc.* A 279, 189–218.

Kidson, C. and Heyworth, A. (1976). The Quaternary deposits of the Somerset Levels. *Q. Jl. Eng. Geol.* 9, 217–35.

Kidson, C. and Wood, R. (1974). The Pleistocene stratigraphy of Barnstaple Bay. *Proc. Geol. Ass.* 85, 223–38.

McMillan, N.F. (1938). On an occurrence of Pliocene shells in Co. Wicklow. *Proc. Lpool. geol. Soc.* 17, 255–66.

Mitchell, G.F. (1972). Further excavations in the early kitchen-midden at Sutton, Co. Dublin. *J. Roy. Soc. Ant. Irel.* 102, 151–59.

Mitchell, G.F. (1976). *The Irish Landscape.* Collins, London.

Mitchell, G.F., Catt, J.A., Weir, A.H., McMillan, Nora F., Margerel, J.P. and Whatley, R.C. (1973). The Late Pliocene marine formation at St. Erth, Cornwall. *Phil. Trans. R. Soc.* B 266, 1–37.

Mitchell, G.F. and Stephens, N. (1974). Is there evidence for a Holocene sea-level higher than that of today on the coasts of Ireland? *Coll. Internat. C.N.R.S.* 219, 115–25.

Monnier, J.-L. (1974). Les Depôts Pléistocènes de la Region de Saint-Brieuc. *Bull. Soc. géol. minér. Bretagne.* C 6, 43–62.

Morzadec-Kerfourn, M.-T. (1974). Variations de la ligne de rivage Armoricaine au Quaternaire. *Mém. Soc. géol. minér. Bretagne.* 17, 1–208.

Morzadec-Kerfourn, M.-T. (1975). Le Plio-Quaternaire marin de Pont-Rouz (Côtes-du-Nord) *C. r. hebd. Séanc. Acad. Sci., Paris.* 280, ser.D.

Peacock, J.D. (1974). Borehole evidence for late- and post-glacial events in the Cromarty Firth, Scotland. *Bull. geol. Surv. Gt. Br.* 48, 55–67.

Pennington, W. (1975). A chronostratigraphic comparison of Late-Weichselian and Late-Devensian subdivisions, illustrated by two radiocarbon-dated profiles from western Britain. *Boreas* 4, 157–71.

Prior, D.B., and Holland, S. (1975). A Late Quaternary Sequence at Carnlough, Co. Antrim. *Ir. Quat. Res. Meeting, Coleraine,* 4–5.

Shackleton, N.J. and Opdyke, N.D. (1973). Oxygen isotope and palaeomagnetic stratigraphy of Equatorial Pacific Core V28–238: oxygen isotope temperatures and ice volumes on a 10^5 year and 10^6 year scale. *Quaternary Res.* 3, 39–55.

Sissons, J.B. (1974). The Quaternary in Scotland: a review. *Scott. J. Geol.* 10, 311–38.

Sparks, B.W., and West, R.G. (1972). *The Ice Age in Britain,* Methuen, London.

Thom, B.G. and Chappell, J. (1977). Holocene sea-level change: an interpretation. *Phil. Trans. R. Soc.*

Tooley, M.J. (1974). Sea-level changes during the last 9000 years in north-west England. *Geogrl. J.* 140, 18–42.

West, R.G. (1968). *Pleistocene geology and biology.* Longmans, London.

West, R.G., and Wilson, D. Gay (1968). Plant remains from the Corton Beds at Lowestoft, Suffolk. *Geol. Mag.* 105, 116–23.

Zagwijn, W.H. (1974). The Palaeogeographic evolution of the Netherlands during the Quaternary. *Geologie. Mijnb.* 53, 369–85.

Zagwijn, W.H. (1975). Variations in climate as shown by pollen analysis, especially in the Lower Pleistocene of Europe. *Ice Ages: ancient and modern Geol. J. Spec. Issue* (ed. A.E. Wright, and F. Moseley), 6, 137–52.

Zagwijn, W.H., and van Staalduinen, C.J. *Toelichting bij Geologische Overzichtskaarten van Nederland.* Rijks Geol. Dienst, Haarlem.

14

The Quaternary of the North Sea

I.N. McCAVE, V.N.D. CASTON, and N.G.T. FANNIN

Abstract

The Quaternary deposits of the North Sea reach a thickness of 1000 m in the Central Graben, suggesting a continuing tectonic control of deposition. Sedimentation rates are up to ten times those for the Tertiary. The morphology of the area includes a series of banks along the Norwegian trough interpreted as moraines, deep scours of possible 'tunnel valley' origin, and the Dogger Bank, an accumulation of fluvioglacial outwash. In the central North Sea borings and seismic work have revealed a succession which appears to be mainly Weichselian in age, both marine and glacial deposits being present. By contrast the southern North Sea contains a relatively complete sequence from the Elsterian to the present. Modern deposition involves sand wave and sand bank movement in the southern North Sea and mud deposition in the Norwegian Trough.

Introduction

The Quaternary deposits of the North Sea, probably over 400 000 km² in extent and up to 1000 m thick, are poorly understood. There are few boreholes and the submarine topography is difficult to interpret.

Until only a few years ago the area was given an involved glacial history based on the slender evidence of morphology with some sediment data, and extrapolation from positions of inferred ice-fronts on land. The latest of this type of reconstruction is Reinhard's (1974) work. Now more substantial data are beginning to appear (e.g. Oele 1969, 1971a, b; Fisher *et al.* 1969; Eden *et al.* 1977), but radiocarbon and palaeontological datings are too few to establish a reliable regional stratigraphy. For that reason this chapter will treat the deposits areally rather than stratigraphically.

Methods

Apart from the normal methods for analysis of Quaternary materials, the marine geologist has available some additional techniques. These include echo sounding, side-scan sonar mapping, continuous reflection profiling, and various marine methods of coring and drilling.

Thickness of Quaternary deposits

An isopach map of total Quaternary thicknesses in the North Sea has been constructed from information obtained from over 200 wells drilled in all five national sectors, supplemented by seismic reflection profiles and published work (Caston 1977a, b).

Information concerning the thickness in these wells (Fig. 14.1) has been provided by the operators concerned; no re-examination of the basic data has been undertaken. Identification of the base of the Quaternary depends upon a number of factors, which in the southern North Sea include lithological changes and their associated electric log responses. The principal criteria have, however, been palaeontological.

In the southern North Sea identifications of the Pleistocene and, where present, the Pliocene, has commonly been based upon assemblages of benthonic Foraminifera and, to a lesser extent, ostracods, using zonations defined primarily in the Netherlands. Extrapolation from the land has provided satisfactory results; in addition, lithological changes, especially in the south and west, where Pleistocene sediments overstep the Tertiary and rest directly upon Mesozoic strata, enable positive determinations to be made.

Moving northwards in the deeper Tertiary basin, fine-grained Pliocene sediments pass virtually unchanged into the Pleistocene,

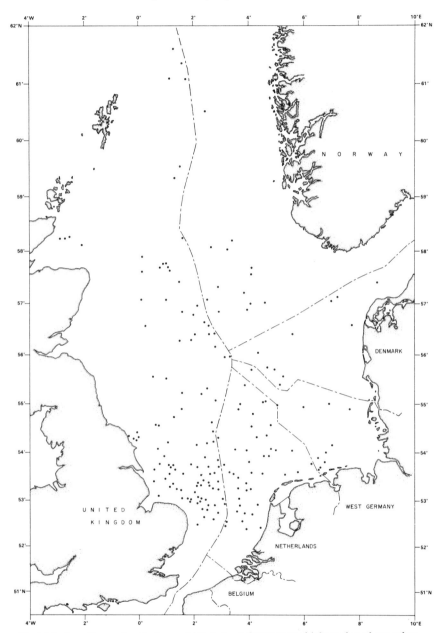

Fig. 14.1. Locations of wells from which data on Quaternary thickness have been taken.

providing few obvious lithological markers (Caston 1977*b*). Distinct
faunal assemblages are also less easy to identify, possibly because of
the earlier influence of cold-water conditions with increasing latitude.

Also because of poor sampling, many operators tend not to distinguish between Pliocene and Pleistocene and hence the isopachs of the northern area, especially north of 58°N. should be treated with considerable caution.

The resultant isopach map (Fig. 14.2) has been prepared principally from new well data but it also incorporates published work listed in the references.

The most conspicuous feature of the map is the trough of thick Quaternary sediments trending N.N.W. from the West Frisians to about latitude 57°, with two closed basins where thicknesses exceed 1000 m. It roughly overlies the Central Graben, the major tectonic feature of Mesozoic and Cainozoic age (Kent 1975). At the south end another basin, offset to the west, is a continuation of the grabens and horsts in the Netherlands Quaternary. North of latitude 57° information becomes much more sparse. For considerable distances off the British coast, from latitude 54° right up to Shetland, sediments rarely exceed 100 m and often are absent. On the east side of the North Sea, inadequate data suggest only a gradual increase in thickness west of the German and Danish coasts until the Central Graben is reached. Off the west coast of Norway there is a linear accumulation of over 400 m of Quaternary corresponding approximately with the Viking Graben, then a zone in which the Quaternary is locally less than 100 m thick, comparable with the Vestland–Stavanger Ridge on Kent's (1975) map, and finally another linear Quaternary deep, which corresponds to the northern part of the Bergen Basin.

The relatively close correlation between the distribution of thick Quaternary deposits and the underlying Mesozoic/Tertiary tectonic features suggests that structural control of sedimentation has continued virtually up to the present day. In the thicker developments the average sedimentation rate may have been as high as 0·3 to 0·5 m per 1000 years, which is up to ten times that for the Tertiary.

Morphology and gravels, valleys, and moraines
Gross morphology (Fig. 14.3.)

The North Sea is very shallow (mainly <50 m) south of the Dogger Bank which extends north-eastwards from the Yorkshire coast. North of the Dogger Bank the North Sea deepens steadily to about 180 m (100 fathoms) north-east of Shetland. The main departure from a planar, even-deepening morphology is a broad depression centred on 58°–59°N. and 0° to 1°E. together with a positive series of banks running north–south from 2°E. at 61°N. to 3°E. at 58°N. (Flinn 1973, Fig. 3). The other major feature of the North Sea is the

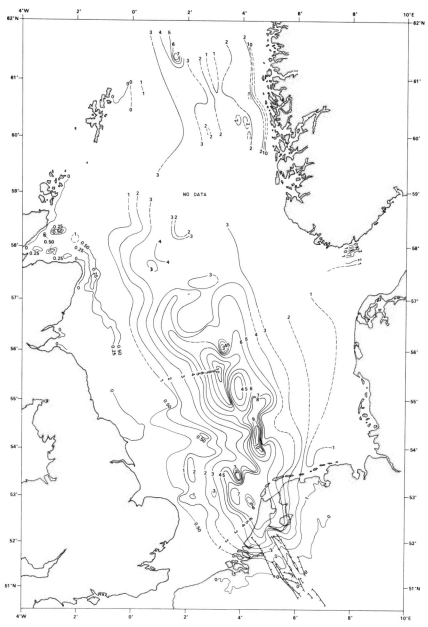

Fig. 14. 2. Isopachs of the Quaternary in hundreds of metres

Norwegian Trough. This is mainly 250 to 300 m deep west of Stavanger and Bergen, but in the Skaggerak south of Oslo reaches 700 m.

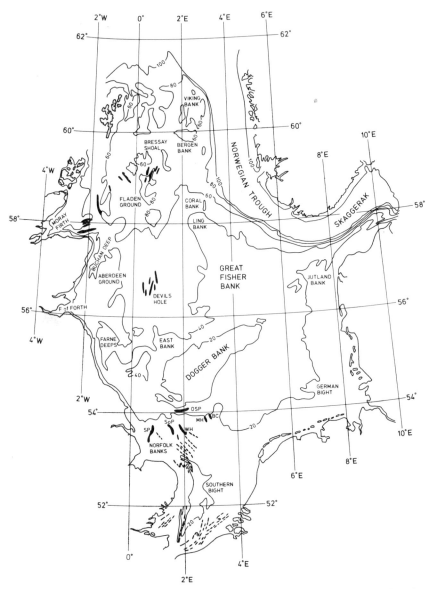

Fig. 14.3. Outline bathymetry of the North Sea, depths in fathoms (1 fathom =1.83 m). Solid black areas indicate deep valleys. Letters by these in the southern area are SP — Silver Pit, SoP—Sole Pit, WH—Well Hole, OSP—Outer Silver Pit, MH—Markhams Hole, BC—Botney Cut. Dashed lines in the southern North Sea indicate crests of sand banks.

Gravels and moraines

Many of the gravel patches in the North Sea have been interpreted as moraines, notably by Pratje (1951). He drew three lines joining gravels and/or banks which he continued on land as the terminal moraines of the Warthe, Frankfurt, and Pomeranian Stadials. The line given for a Warthe moraine could easily have been given many alternative positions based on other gravel patches and it seems unwise to perpetuate it. Taking the supposed Frankfurt and Pomeranian moraines together as late Weichselian Moraines (the evidence does not allow separate trends to be convincingly drawn for the two), they correspond to the positive series of banks running along the western and southern margin of the Norwegian Trough mentioned above.

The profiles of Sellevol and Sundvor (1974, Fig. 6) show that the sub-Quaternary surface displays a shallow depression, deepening towards Norway, attributed to glacial erosion. It forms a part of the Fennoscandian 'arc of exhumation', an erosional feature found concentric with and inside the margin of the ice-sheet and recognized from many other places by White (1972), Holtedahl (1958), and Shepard (1931). The Quaternary sediments over this surface are much thicker in the area of the late Weichselian Moraines (>300m) than in the trough itself (70 to 250m) and this accentuates the form of the depression. While on morphological grounds it seems likely that surface sediments and topography along the margins of the trough may be late Weichselian, it is unlikely that the whole 300 m thick pile beneath is of the same age.

The date of erosion of the basal sub-Quaternary surface sloping down towards Norway is also unknown. However, given the existence of this slope, it was probably the cause of glacial stagnation leading to thicker sediment deposits along the margin of the trough. We suggest that some initial erosional event scoured out a depression and subsequently ice movement slowed down over the outer slope of this depression, yielding a thicker deposit on its margin and thereby accentuating its depth and form.

There has been little detailed petrography of the North Sea gravels other than the work of Veenstra (1969). He recognizes a number of associations. Sandstone, limestone, and prophyry of British origin are found off the coast of Lincolnshire and Yorkshire, on and to the south of the Dogger Bank. Prophyry, granite, flint, sandstone, and limestone of supposed Scandinavian origin are found on Cleaver Bank to the south of the Dogger Bank. All these materials are assigned by Veenstra to Weichselian Moraines. Off the Dutch island of Texel are also found Scandinavian erratics, this time of Saale Moraines. An extensive area of flint, limestone, and quartzite gravel occurs off the East Anglian Coast and in the Hinder Banks region off the Belgian coast. The composition differs from that of the Rhine and Meuse, is mainly of Cretaceous and Jurassic material, and is thought to be of fluvial origin, partly from the Ardennes and partly from the Thames. Well-rounded pebbles with percussion marks suggest that some may have been reworked on beaches during the Flandrian transgression.

Valleys
The North Sea floor carries numerous valleys with dimensions of the order of 10-50 km long, 1-4 km wide, and relief of 20-150 m but mainly of 50-100 m (Table 14.1). They have been described by Valentin (1955), Flinn (1967, 1973), Donovan (1973) and D'Olier (1975) and suggested to be 'tunnel valleys' like the 'Rinnentaler' of Denmark. These features are probably mostly glacial-related but whether they all have the same origin or are all glacial is doubtful. Similarly shaped depressions have been found on non-glaciated shelves with strong tidal currents and they also occur in some tidal inlets between barrier islands.

Features of comparable scale to these valleys, infilled with sediments, have been detected by continuous seismic profiling (Dingle 1971; Holmes 1977; Caston 1977b; see next section). They apparently represent two periods of cutting, one pre-late Weichselian glacial, the other postglacial or late glacial, the Lower Channels and Upper Channels respectively. The distribution of these filled channels is more extensive than the unfilled ones and, if taken together, no simple pattern emerges such as that on which Flinn (1967) postulated an arcuate ice front and lake in the central northern North Sea. Many questions remain, notably whether the filled and unfilled channels are basically the same but only some are filled, the mode(s) of origin of the features, and the times of cutting and filling. The relation of these features to an ice-front is most uncertain in the case of Flinn's (1967) postulate, somewhat better documented in the case of the southern lobe of the East English Glacier of Valentin (1955) because of additional evidence on land, and a subject for debate in the case of D'Olier's valleys as it has not generally been thought that the outer Thames estuary was ice covered. It should be noted here that one set of Lower Pleistocene filled channels which are thought to be not glacially controlled are those discussed by Funnell (1972) beneath the crags of East Anglia.

The central North Sea succession
Seismic and drilling data suggest up to 500 m of Quaternary sediments in the area of the central trough in the northern North Sea.

Table 14.1.

Dimensions of deep valleys in the North Sea

Area	Relief (m)	Max depth below MSL (m)	Width (km)	Length (km)
A. Unfilled				
N. Fladen Ground[1]	20-100	225	1-4	< 35
Buchan Deeps[1]	50-150	248	1-3	< 50
Aberdeen Deeps[1]	37- 50	131	2-3	< 50
Devils Hole group[1]	50-150	237	~ 1	> 35
Silver Pit group[2]	30- 70	95	0·5-3	9-30
Thames Estuary[3]	~ 30	60	1-4	2·4-23
Hurd Deep[4]	100	175	5-6·5	165
Other English Channel[4]	25- 30	90-110	~ 2	15-30
B. Filled				
East Bank[5]	30-150	250	1-5	> 25
Lower Channels	< 125	280	< 2	~ 9?
Upper Channels	10- 50	250	?	?
Crag depressions[6]	30- 50	—	3-7	15-30

[1] Flinn (1967, 1973);   [2] Donovan (1973);   [3] D'Olier (1975);   [4] Boillot (1963);
[5] Dingle (1971);   [6] Funnell (1972).

They thin abruptly along a line which curves sinuously northwards between 0° and 2°E. from 56°N. to 62°N., and wedge out towards the coast. This thinning, combined with the rapid facies variations typical of glacial sediments, causes problems in relating the inshore and offshore successions. A generalized stratigraphy which links the successions is shown diagrammatically in Fig. 14.4. It has been established from the IGS reconnaissance programme in the area of the British sector of the North Sea between 56° and 58°N., in the Moray Firth west of 2°W. and in unlinked areas around Orkney and Shetland and the Brent to Thistle oil fields (southern half of British Block 211) supplemented by intensive studies in the Forties area (Caston 1977*b*) and in a small area around 58°5'N., 0°35'E. (Holmes *et al.* 1975; Holmes 1977). The succession established is based mainly on shallow seismic data (e.g. Fig. 14.5) but is supplemented by sea-bed sampling and shallow drilling. The work is still in progress, and local names describing groups of beds are used informally.

The relations between the stratigraphic units are shown schematically on Fig. 14.4. Notes on these units appear below.

Basal Beds. Apparently conformable with Upper Tertiary and bounded above by a disconformity. Thickness 50 to 100 m. Seismically poorly bedded, possibly gas-bearing.

Aberdeen Ground Beds. Sharp disconformity at the base and erosional, channelled top with relief of over 100 m. Thickness over 200 m in places. Seismically well bedded. Upper 100 m comprise car-

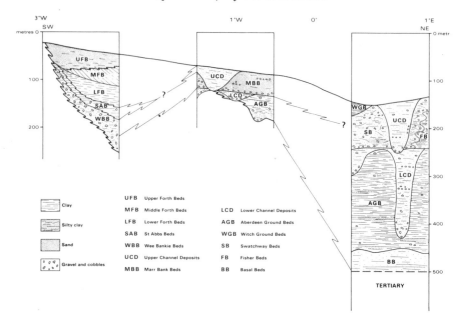

Fig. 14.4. Diagrammatic composite sections of the Quaternary succession of the central North Sea. The sections are based on a combination of drilling and seismic data (Holmes *et al.* 1975; Chesher and Lawson 1976; Thomson and Eden 1976). Seismic velocities used are 1600 m/s for Witch Ground Beds, and 1800 m/s for all other sediment layers.

bonaceous dark grey sandy clays with thin sands, marine bivalves, and a cold water micro-fauna and flora. Radiocarbon dates from partially lignitized wood chippings recovered during drilling at 56° 34′N. 2° 32′E. are 23 170 BP (at 109–127 m sub-bottom), 32 682 ± 530 BP (at 246–254 m), and 47 716 ± 1940 BP (at 307–315 m). Yew, juniper, spruce, birch, alder, poplar, and willow have been identified. The dates and flora indicate mid-Weichselian and interstadial conditions with the yew, poplar, and willow probably having drifted in from the south. Thus the bulk of the Aberdeen Ground Beds appear to be marine and mid-Weichselian in age.

Lower Channel Deposits. Depressions in the Aberdeen Ground Beds are filled with sands, gravels, and some interbedded clays. No regional orientation pattern is yet revealed.

Swatchway Beds and eastward laterally equivalent *Fisher Beds.* Overlie the lower channel deposits with complex, laterally discontinuous, variable lithologies including clays, pebbly clays, sands, and gravels, all probably of glacial origin. Thickness about 100 m. Comprise layers B to D in the Forties area (Fig. 14.5). Irregularly bedded with much evidence of cut and fill. Thought to be of glacial origin.

Upper Channel Deposits. Channels cut into the Swatchway Beds are fiꞮed with sands and gravels.

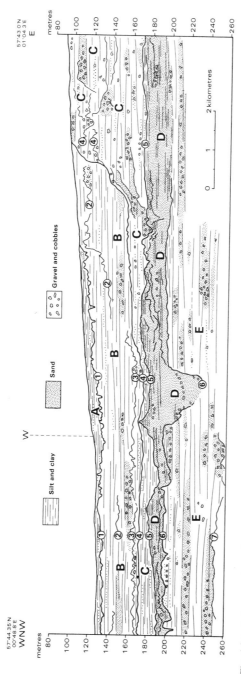

Fig. 14.5. Interpretation of constant scale profile of 1 kJ sparker line across the Forties Field, (57° 44·3'N, 0° 48·8'E. to 57° 43'N., 1° 4·4' E.) based on Caston (1976a, Fig. 5). The vertical exaggeration is 1:25, and layers A to E and reflectors 1 to 7 follow Caston's nomenclature. The thick numbered lines are correlatable reflectors and the thin lines refer to discontinuous minor reflectors. Nominal depths are calculated using velocities of 1478 m/s in water, and 1676 m/s in all sediment layers. The complex of beds labelled C on the right of the diagram are thought to be a late Weichselian moraine.

Witch Ground Beds. Soft, dark grey, carbonaceous silty clays of probable Flandrian age filling broad depressions in the surface of the Swatchway Beds (layer A, Fig. 14.5).

In the inshore area the Aberdeen Ground Beds are cut by channels which are themselves truncated by an extensive eastward-dipping erosion surface (Thomson *et al.* 1977).

Marr Bank Beds. Up to 40 m of shelly and gravelly marine sands and clays overlying the erosion plane. Partly equivalent to the Swatchway Beds. Upper surface channelled and filled with upper channel deposits.

Wee Bankie Beds. Tills preserved in hollows in bedrock, equivalent to Marr Bank Beds (in part?).

St. Abbs Beds. Dark brown clays with arctic fauna, about 10 m thick. Probable equivalent of the Errol Clay aged $\cong$ 13 000 years.

Forth Beds. Up to 100 m of clays coarsening upwards to silty clays and sands. The base of the Flandrian is somewhere in the middle Forth Beds.

The general succession is thus marine deposits over a disconformity, these in turn overlain by a glacial sequence. Going towards the land (westwards) the marine deposits are cut out and till lies directly on pre-Quaternary. The whole sequence appears to be Weichselian with some Lower Pleistocene(?) deposits below the basal disconformity.

Areas peripheral to the central North Sea
Further details have been obtained from the peripheral areas mentioned below.

Moray Firth
Two separate areas of Quaternary sediments are recognized here (Chesher and Lawson *in preparation*). In the nothern part the Quaternary rests on a platform of Upper Palaeozoic and Mesozoic rocks, while to the south the sediments are found in a series of east–west basins. The basins were carved by converging ice moving out of Cromarty Firth and the Great Glen and the complexity of their infills has so far prevented inter-basin correlation. The general succession in both areas is of basal till interbedded with local glacio-lucustrine deposits, followed by a few metres of pinkish or brownish clays with an arctic fauna, probably equivalent to the St. Abbs Beds and the Errol Clay. Above these clays are a series of greenish clays and sands of Flandrian age, equivalent in part to the Forth Beds and the Witch Ground Beds. The basal tills are therefore probably equivalent to the Swatchway Beds and the Wee Bankie Beds.

The Norwegian Trough

In a recent seismic reflection study van Weering *et al.* (1973) recognized four principal units in the Norwegian Trough down to a depth of about 100 m sub-bottom. The uppermost two units are thought to be postglacial infill and the lower units, which have an irregular surface of high reflectivity, to be tills of probable Weichselian age. Flinn (1967, 1973) suggested that the Fladen Ground was ice-free in the Weichselian but van Weering *et al.* find glacial drift units on the western side of the Trough at the latitude of the Fladen Ground and it appears that glacial drift persists through the latter area beneath marine sediments.

The Frigg Field and booster station

Borings in the Quaternary sediments have been made to 100 m and the sediments and fauna described from the upper 50 m at the Frigg Field (59°53′N., 2°4′E.) and the upper 35 m at the pipeline booster station site (58°50′N., 0°17′W.) (Løfoldli 1973). Five foraminiferal assemblages were recognized at Frigg, the upper two being recent and postglacial, the third possibly from the last glacial maximum on the basis of its arctic faunas, and the lower two zones supposedly interstadial, pre-late Weichselian. Although the succession contains gravels, there is little clay or silt, no boulder clay, and no unequivocal glacial deposits, and there are marine faunas throughout. This poses problems in relation to supposed late Weichselian Tills recognized in the central North Sea and the glacial deposits traced seismically by van Weering *et al.* (1973). The deposits at the booster station (0–35 m) contain the upper three zones, pre-last glacial sediments not being encountered. It may be that some part of the northern North Sea was ice-free in the late Weichselian glacial advance, but this is difficult to reconcile with the suggestion of Hoppe (1972, 1974) that Shetland was glaciated in the Weichselian by ice which came from the east, i.e. probably from Scandinavia.

The Dogger Bank and southern North Sea

In contrast to the areas further north, the southern North Sea has a relatively complete suite of deposits from the Elsterian/Anglian onwards. The account given here is largely based on the work of Oele (1969, 1971*a*, *b*) who undertook a survey with a vibrocorer. However, this does not give penetration greater than 10 m, and the amount of published reflection seismic data is very small.

Beneath the Dogger Bank is a layer of Elsterian fluvioglacial clay 30 m thick overlying unfossiliferous sands. No Holsteinian deposits have been found in the area of the Bank, but to the south west in

Silver Pit, Fisher *et al.* (1969) report marine Hoxnian. In the Southern Bight a boring at 52°45′N., 3°25′E. encountered 7 m of marine Holsteinian, and elsewhere over 20 m has been found. The Saalian Glacial was responsible for boulder clays in a trough to the south of the Dogger Bank, and in the Southern Bight Saalian moraines of the Drenthe Formation are recorded as far south as 52°20′N. off the Dutch coast. Extensive Eemian deposits of both fresh water (the Brown Bank bed) and marine origins are found in the Southern Bight. Further north, Eemian deposits are known only from a subsurface valley east of 4°20′E. and are not known from the British side at all. The Weichselian glaciers of Scandinavian origin appear to have stopped at the north side of the Dogger Bank where the Elsterian clays show an ice-pushed ridge. The bank itself is an accumulation of fluvioglacial outwash sands overlain by coversands of the late Weichselian and veneered with Holocene sands some of which are reworked coversands containing mammal bones (Oele 1971*a*, Fig. 3; Stride 1959). Off the mouth of the Rhine, fluvial Weichselian deposits of the Kreftenhaye Formation occur, while further north their equivalents are wind-blown sands of the Twente Formation.

Because there have been few deep borings and little sparker work in the area, we do not have much information on the Lower Pleistocene. The situation on land at depth in Holland is summarized by de Jong (1967) and in East Anglia by Funnell (1972) (see also Chapter 18). At sea there is no information from the British sector on any part of the section save that on thickness given above, and the identification of Hoxnian deposits and late Devensian Purple Till in the Silver Pit (Fisher *et al.* 1969; Donovan 1973).

The Holocene

In the southern part of the North Sea, Oele (1969, 1971) records a sequence very similar to that in Holland, namely a Preboreal to Boreal lower peat overlain by clays of fluvial to brackish-water origin, tidal flat deposits, and sands known as the Elbow/Calais deposits. The Sub-boreal and Atlantic are probably unrepresented and the early Holocene sediments are capped by young (Subatlantic) marine sands. The position of submarine peats fits in with Jelgersma's (1961) sea-level rise curve. This curve was employed by D'Olier (1972) together with results of seismic reflection profiling to illustrate the Holocene flooding of the Thames Estuary between 9600 BP and 8000 BP. He finds that over the last 9000 years, 0·27 km^3 per century of sediment has been introduced into the estuary.

Elsewhere, south of the IGS central North Sea area, there are only fragments of information available on the Holocene. Eisma (personal communication) has examined the deep scours of the Silver Pit region with a continuous reflection profiler. He notes that the basins nearer the coast, Silver Pit, Sole Pit, and Well Hole have negligible Holocene infilling, probably because of tidal current scour whereas the outermost hollows, Markhams Hole and Botney Cut in an area of slower tidal currents, have accumulated up to 50 m of fill. Veenstra (1965, 1969) finds finer gravel dominant in samples from greater depths on the south side of the Dogger Bank with coarser gravel higher up and surmises that this effect was produced by wave winnowing of glacial gravels at lower sea-level.

On the north-west side of the Dogger Bank in the East Bank area, Dingle (1971) finds the surface sediments formed into elongate banks up to 30 m in thickness rising to 40–50 m below sea-level. The length (up to 25 km) and spacing (order of 4 km) of these banks are similar to the tidal sand ridges of the Southern Bight (Houbolt 1968), suggesting a similar tidal origin. On this hypothesis the ridges would have formed at a time when relative sea level was about 40 m lower than at present. A short way to the W.S.W. of these ridges is an active sand wave field from 0°20′E. to 1°20′E. and south of 54°30′ (Dingle 1965). This juxtaposition of sand banks and sand waves is similar to that found in the Southern Bight with the Hinder banks to the south of the sand wave field, but there both the banks and sand waves are active whereas the banks north west of the Dogger Bank are probably not.

In the Norwegian Trough, Unit 1 of van Weering *et al.* (1973) is up to 30 m thick and probably is fine sediment of Holocene age deposited mainly from suspension. Deposition probably continues at the present day, with mud from fluvial and coastal erosion sources being ultimately trapped in the Trough (McCave 1973).

In the central North Sea the Forth Beds mentioned previously are partly Holocene in age. They comprise a silt clay facies and a sandy facies. Away from the Firth of Forth the Holocene is represented by the Witch Ground Beds, which in the Forties area are some sand lenses (layer A, Fig. 15.5) overlain by a layer of sand (Caston 1977b). The acoustically transparent deposits up to 18 m thick recorded by Flinn (1967) in the Fladen Ground area (58°–59°N, 0° to 1°E.) also probably contain Witch Ground Beds and may be equivalent to unit 1 of van Weering *et al.* 1973.

Conclusions

The Quaternary of the North Sea is poorly known at present. Available information suggests that the bulk of the Quaternary of the northern North Sea is of Weichselian age. A preliminary stratigraphic scheme for the northern area has been drawn up by the IGS which promises to be applicable over a wide area. Thicknesses range up to 1000 m in the central trough. From the Dogger Bank southwards a more complete sequence including Elsterian, Holsteinian and Hoxnian, Saalian, Eemian, and Weichselian deposits has been investigated almost entirely in Dutch waters. There remains considerable scope for work in the British half of the southern area of the North Sea.

Acknowledgements

I.N.McC. thanks his colleagues Geoffrey Boulton and Brian Funnell for help in discussion and review of this paper. V.N.D.C. acknowledges permission to publish from the British Petroleum Company Ltd. N.G.T.F. acknowledges the help and permission given by a number of his colleagues to incorporate their data. He publishes with permission of the director of the Institute of Geological Sciences.

References

Boillot, G. (1963). Sur une nouvelie fosse de la Manche occidentale, la fosse du Pluteus. Sur la fosse de la Hague (Manche centrale). Sur la fosse centrale de la Manche. *C. r. hebd. Séanc. Acad. Sci., Paris.* 257, 3448-51, 3963-66, 4199-202.

Caston, V.N.D. (1977a). The Quaternary deposits of the central North Sea, 1 and 2. pp. 1-8. A new isopachyte map of the Quaternary of the North Sea. *Rep. Inst. Geol. Sci.* 77/11, 22 pp.

Caston, V.N.D. (1977b). The Quaternary deposits of the central North Sea, 1 and 2. pp. 9-22. The Quaternary deposits of the Forties field, northern North Sea. *Rep. Inst. Geol. Sci.* 77/11, 22 pp.

Chesher, J.A. and Lawson, D. (in preparation). Marine geology of the North Sea. *Rep. Inst. Geol. Sci.*

Dingle, R.V. (1965). Sand waves in the North Sea mapped by continuous reflection profiling. *Mar. Geol.* 3, 391-400.

Dingle, R.V. (1971). Buried tunnel valleys off the Northumberland coast, western North Sea. *Geologie Mijnb.* 50, 679-86.

D'Olier, B. (1972). Subsidence and sea-level rise in the Thames Estuary. *Phil. Trans. R. Soc.* A272, 121-30.

D'Olier, B. (1975). Tunnell Valleys and associated features of the Southern Bight of the North Sea. *Quaternary Newsletter* 17, 5.

Donovan, D.T. (1973). The geology and origin of Silver Pit and other closed basins in the North Sea. *Proc. Yorks. geol. Soc.* 39, 267-93.

Eden, R.A., Holmes, R. and Fannin, N.G.T. (1977). The Quaternary deposits of the central North Sea, 6. Depositional environment of offshore Quaternary deposits of the Continental Shelf around Scotland. *Rep. Inst. Geol. Sci.* 77/15.

Fisher, M.J., Funnell, B.M. and West, R.G. (1969). Foraminifera and pollen from a marine interglacial deposit in the western North Sea. *Proc. Yorks. geol. Soc.* 37, 311-20.

Flinn, D. (1967). Ice front in the North Sea. *Nature, Lond.* 215, 1151-14.

Flinn, D. (1973). The topography of the sea floor around Orkney and Shetland and in the northern North Sea. *J. geol. Soc. Lond.* 129, 39-59.

Funnell, B.M. (1972). The history of the North Sea. *Bulletin of the Geological Society of Norfolk*, 21, 2-10.

Holmes, R. (1977) The Quaternary deposits of the central North Sea, 5. The Quaternary geology of the UK sector of the North Sea between 56° and 58°N. *Rep. Inst. Geol. Sci.* 77/14.

Holmes, R., Fannin, N.G.T. and Tully, M.C. (1975). Geological report on the Forties pockmark detailed survey area, M.V. Sea Lab. drill sites, Forties to Piper and Forties to Engineering Study Area reconnaissance lines. *Institute of Geological Sciences, Continental Shelf Unit North, Internal Report*, 75/13.

Holtedahl, H. (1958). Some remarks on geomorphology of continental shelves off Norway, Labrador, and south-east Alaska. *J. Geol.* 66, 461-71.

Hoppe, G. (1972). Ice sheets around the Norwegian Sea during the Würm glaciation. In: The Norwegian Sea Region. *Ambio, Special Report*, 2, 25-9.

Hoppe, G. (1974). The glacial history of the Shetland Isles. *Spec. publ. Inst. Geogr.* 7, 197-210.

Houbolt, J.J.H.C. (1968). Recent sediments in the Southern Bight of the North Sea. *Geologie Mijnb.* 47, 245-73.

Jelgersma, S. (1961). Holocene sea-level changes in the Netherlands. *Meded. geol. Sticht.* Series C, 6.

de Jong, J.D. (1967). The Quaternary of the Netherlands. In *The Quaternary*, Vol. 2, (ed. K. Rankama), pp. 301-426. Interscience, New York.

Kent, P.E. (1975). Review of North Sea Basin development. *J. geol. Soc. Lond.* 131, 435-68.

Løfoldli, M. (1973). Foraminiferal biostratigraphy of late Quaternary deposits from the Frigg Field and booster station. *Norges Teknisk-Naturvitenskapelige Forskningsråd, Continental Shelf Division, Report.* 18.

McCave, I.N. (1973). Mud in the North Sea. In *North Sea Science* (ed. E.D. Goldberg), pp. 75-100. MIT Press, Cambridge, Mass.

Oele, E. (1968). The Quaternary geology of the Dutch part of the North Sea, north of the Frisian Isles. *Geologie Mijnb.* 48, 467-80.

Oele, E. (1971a). Late Quaternary geology of the North Sea south-east of the Dogger Bank. *Institute of Geological Sciences, Report*, 70/15, 29-34.

Oele, E. (1971b). The Quaternary geology of the southern area of the Dutch part of the North Sea. *Geologie Mijnb.* 50, 461-74.

Pratje, O. (1951). Die Deutung der Steingrunde in der Nordsee als Endmoränen. *D. hydrogr. Z.* 4, 106-14.

Reinhard, H. (1974). Genese des Nordseeraumes im Quartär. *Fennia*, 129.

Sellevol, M. and Sundvor, E. (1974). The origin of the Norwegian Channel—a discussion based on seismic measurements. *Can. J. Earth Sci.* 11, 224-31.

Shephard, F.P. (1931). Glacial troughs of the Continental Shelves. *J. Geol.* 39, 345-60.

Stride, A.H. (1959). On the origin of the Dogger Bank, in the North Sea. *Geol. Mag.* 96, 33-44.

Thomson, M.E., Moira, E. and Eden, R.A. (1977). The Quaternary deposits of the central North Sea, 3. Quaternary sequence in the west central North Sea. *Rep. Inst. Geol. Sci.* 77/12.

Valentin, H. (1955). Die Grenze der letzten Vereisung im Nordseeraum. *Verhandlungen der Deutschen Geografentag, Hamburg.* 30, 359-66.

Veenstra, H.J. (1965). Geology of the Dogger Bank area, North Sea. *Mar. Geol.* 3, 245-62.
Veenstra, H.J. (1969). Gravels of the southern North Sea. *Mar. Geol.* 7, 443-64.
van Weering, T., Jansen, J.H.F. and Eisma, D. (1973). Accoustic reflection pro-files of the Norwegian channel between Oslo and Bergen. *Neth. J. Sea. Res.* 6, 241-63.
White, W.A. (1972). Deep erosion by continental ice sheets. *Bull. Geol. Soc. Am.* 83, 1037-56

Late Note

Recently Jansen (1976) has published shallow seismic work from the area north of the Dogger Bank. In the general area of the Fladen Ground three acoustic units are recognized: Witch deposits, overlying Fladen deposits, and Hills deposits. The latter are glacial and glacial marine of late Weichselian age. Thicknesses are up to 15, 20, 25 m respectively. The sequence is cut by tunnel valleys post-dating the Hills deposits. To the north of the Dogger bank is a system of channels possibly of Saalian incision with a complex history of filling with possible Saalian, Eemian, and Weichselian deposits overlain by a thin post-glacial fill equivalent in age to the Witch deposits. Tentative correlations with our central North Sea succession are the Witch deposits with the Witch Ground Beds (and the upper and part of the middle Forth Beds inshore, but not with the St. Abbs Beds as suggested by Jansen), the Fladen and Hills deposits with the Swatchway/Fisher Beds. It may be that the Fladen deposits are equivalent to layer B between reflectors 1 and 4 and the Hills deposits to layers C and D (U. and L. Hills ?) on Fig. 14.5.

Jansen, J.H.F. (1976). Late Pleistocene and Holocene history of the northern North Sea, based on acoustic reflection records. *Netherlands Journal of Sea Research.* 10, 1-43.

15

Periglaciation

P. WORSLEY

Abstract

A selective review of recent periglacial work is presented. Studies of mass movement effects have advanced via a geotechnical approach which has identified the frequent occurrence of basal shear surfaces. Large-scale non-diastrophic structures and tors now appear to be diagnostic periglacial features. Patterned ground of several kinds has a wide distribution and is not restricted by any glacial limits. The role of niveo-fluviatile processes is becoming increasingly evident. Permafrost structures are known from five glacial stages and during several periods within the last of these. Only a limited amount of periglacial activity persists today and the same applied during the Flandrian.

Introduction

Periglacial studies in the British Isles have a long history of investigation, for the products of periglacial processes are ubiquitous throughout the region. The earliest geological surveyors were especially impressed by the effects of former mass movements which had produced a surficial capping to many of the exposures, and today one is impressed by their accord with modern concepts of periglaciation. This chapter presents a selective review of work completed within the last decade. Particular attention will be devoted to mass movement, certain patterned ground and involution phenomena, and an evaluation of the role of rivers. The various periods within the Pleistocene when permafrost growth occurred will be outlined, concluding with a comment on Flandrian activity. Earlier summaries of an extensive literature with detailed bibliographies are available by FitzPatrick (1956), and Waters (1964). Aeolian processes and their products are specifically excluded since they are considered in Chapter 16.

Slopes and mass movements

Head

The products of periglacial slope processes in general have traditionally been termed 'head'. It is emphasized that head deposits are not envisaged as materials related exclusively to solifluction and the definition is sufficiently loose to include deposits intermediate in character between such deposits and true fluvial sediments. South of the Devensian glacial ice limit the widespread occurrence of surficial head testifies to the importance of periglacially induced mass movements during the last glacial stage and the more limited presence within the glaciated area signifies that the periglacial environment was relatively short-lived after the main ice wastage event.

An appreciation of the primacy of head as a Quaternary *facies* comes largely from the mapping programme of the Geological Survey, and their 'drift' map editions published since the 1930s give ample evidence of this. Mottershead (1971) demonstrated the inter-relationship between head, blockfields, and tors in a head sequence lying over a rock platform on the south Devon coast. Here the head exceeds 33 m in thickness and comparable values are common in many coastal situations, especially along southern coasts. On the west coast of central Wales, Watson and Watson (1967) have re-interpreted what had previously been regarded as a till body over scree banked against a former cliff line, as being entirely head, largely on macrofabric evidence. Fossil block-streams, usually embedded in head materials, are known from several localities in the south, and Williams (1968) quotes a total length in excess of 10 km for a particularly well-developed example in Cambridgeshire. A detailed analysis of a similar feature in Wiltshire by Small *et al.* (1970) concluded that solifluction of the head was the chief mechanism whereby the blocks were transported. The most recent discovery in this field is a fossil rock glacier in north west Scotland (Sissons 1976), identical in its characters with modern active forms. It is possible that other examples within this class of phenomena will be located in the future and it is here suggested that the large complex fan of solifluction material noted by Watson (1966), in front of what he interprets as a nivation cirque in mid-Wales, might be a related form.

Periglacial mudslides

A major advance in the understanding of head genesis has resulted from geotechnical site investigations into the role of periglaciation in determining the form and nature of the surficial cover developed upon clay bedrock. These clays, usually of pre-Quaternary age, are over-consolidated and fissured. Natural outcrops are rare since slopes are invariably masked by a head mantle which may exceed 6 m in thickness and which often consists of an upper structureless layer with clasts of material derived from outcrops upslope, and a lower component of weathered clay derived from the bedrock underneath.

At present, these slopes usually have a convexo-concave profile and a maximum slope between 7 and 10°. Although they do not display any marked surface expression of instability, sections reveal polished and slickensided shear surfaces. Most of these are either continuous shears sub-parallel to the surface slope or discontinuous features which have a tendency to approach the surface downslope. It is clear that all the material above the lowest slip surface at any locality has been subject to downslope mass movement. Since the slopes are generally stable under contemporary conditions in the natural state it seems that in the past the shear strength of the surficial materials must have been significantly reduced. When movement occurred, preferential alignment of the clay particles along the shear surfaces led to the characteristic slip surfaces observed in the field. Such surfaces reduce the residual strength of the slope mantle of head compared with adjacent unsheared material. To re-activate movement the modern drainage state of the mantle must change so that pore water-pressures develop to a level which exceeds that of a ground water-table lying at the surface of the ground. Hence the shear surfaces are interpreted as fossil features and are thought to result when artesian pore water pressures increase to failure condition within either a seasonally frozen or an active layer above permafrost. Approximate undrained conditions would arise when thaw rates were sufficiently rapid relative to the rate of consolidation (Hutchinson 1974). Such mass failure is regarded as a kind of solifluction but, since distinct shear is involved it could be considered either as a variety of slab slide failure or periglacial mudslide (Chandler 1972).

A major difficulty is a lack of data from modern periglacial environments related to analogous geological situations which must have existed in Britain. Identification of the processes responsible for the fossil periglacial mudslides must be based largely upon inference. Following Chandler (1972), three classes of periglacial mudslide may be recognized, each involving the generation of artesian pore pressures:

(i) Sheet movements, entirely or largely composed of clayey sediment. Apart from those slopes greater than $7°-8°$ which are not exclusively periglacial, these occur in cohesive materials on slopes of less than $7°-8°$. The circumstances of pore water pressure generation are obscure.

(ii) Sheet movements of granular permeable materials over an impermeable base, on slopes down to as low as $3°-4°$. These are probably due to extensive slow freezing of the ground surface and do not necessitate the presence of permafrost.

(iii) Individual 'successive emergent' movements in cohesive materials—mainly clayey sediments. Again these characterize slopes down to $3°-4°$ and seem to be due to local variations in permeability inhibiting free drainage through the soil, thereby causing a 'blanketing' effect. Localized excess pore water pressures could operate in conjunction with a surface impermeable layer caused by one of several mechanisms including nocturnal surface freezing, surface silt accumulations from seepages, differential frost shatter with depth and sub-surface sorting.

Non-diastrophic superficial structures

This group of structural deformation features includes cambers, dip and fault structures, sags, and valley bulges. They are widespread throughout southern areas, especially where the pre-Quaternary sequence involves an alternation of massive beds and incompetent materials as in the Carboniferous and Jurassic Systems. Currently in those areas displaying the structures general stability seems to prevail and has done so through at least the Flandrian.

Although there is no agreement as to precisely how the forms were produced, there is a consensus that the primary cause is the creation of differentially loaded systems by erosional processes within a periglacial environment. However, the periglacial connotation has sometimes been invoked with little supporting evidence. Significantly perhaps, one of the authors of the initial paper on the topic (Kellaway 1973) has recently argued that a radical reappraisal is required by suggesting that the structures are intimately related to deglacierization. He postulates that overdeepening of a pre-existing, relatively low relief landscape was caused by a combination of glacial erosion and meltwater downcutting, with the large-scale forms being produced as a response to this process. In this view, the environment remains periglacial in the widest sense but is closely linked to glaciation. However, these ideas necessitate glacial limits which are currently unacceptable to many workers if all the known examples of the features are to be so explained.

Yet the concept of sudden incision is probably basic to the genesis of the structures although the key process might be 'normal' fluvial erosion of valley axial materials. The chalk cliffs of the English south coast expose cross-sections of many truncated dry valleys where the chalk, when traced toward the valley bottoms, becomes progressively disrupted so that directly below the valley axes it has the appearance of head. It is suggested that the growth of ground ice segregations was enhanced beneath the valleys and, as a result, they are now underlain by low strength materials. This mechanism could operate in any fine grained rock which would be rapidly eroded if sufficient gradient were available following climatic amelioration. In this way earlier suggestions of former frost weathering, postglacial rebound of compressed clays, differential loading, and water content of the argillaceous beds could be incorporated into a single hypothesis.

Tors

The investigation of tors seems almost to have been a British prerogative. It is also evident that tors are not always regarded internationally as periglacial forms as is shown by their absence from Washburn's (1973) treatise. Tors have long been recognized in contemporary periglacial regions but the term was not established until Linton's classic paper (1955) in which he argued for a two-stage formative model involving initial differential deep chemical weathering in a tropical climatic environment and a later stage of exhumation of unweathered residuals. An alternative view was advanced which argued that tors were essentially unique periglacial forms produced by frost shatter processes in conjunction with mass movement.

Recently the issue has been pursued by others and most of the discussion has centred upon the granite tors of Dartmoor in southwest England. Eden and Green (1971) showed that the weathered granite (growan) was restricted to the margins of the granite outcrop and the main river valleys and seismic work revealed that the undissected, supposedly Tertiary erosion surfaces had only a thin weathered cover. Although they discarded tropical analogies they did, nevertheless, support a two-stage formative model. Utilizing scanning electron microscope techniques, Doornkamp (1974) tackled the fundamental problem of growan genesis. A comparative study with material from known tropical areas showed that the textures on the detrital quartz from the growan showed little evidence of widespread chemical weathering but rather an origin by mechanical breakdown. Further, the overlying head revealed the same textures. Thus it seems that mechanical processes are the more important in achieving the granular disintegration of the granite, tending to support a periglacial mode of formation in each of the two stages.

Tors are not restricted, however, to granite outcrops and fine groups are developed upon other lithologies including quartzites, dolomitic limestones, and the coarse arkosic sandstone members of the Namurian in northern England. With reference to the latter forms, Linton had speculated that the Pennine valleys should contain sequences of fine grained sediments which had been removed during the exhumation of the tors in accordance with his model. Since then, on the western margins of the Pennines, thick alluvial fan sediments have been recognized and these seem to have been largely derived from the Pennine valleys when periglacial conditions prevailed. It is probable that most of this material originated from the granular disintegration of the sandstones and was directly associated with the revelation of the tors. In conclusion it is clear that although the precise nature of the rock breakdown may still be debatable, extensive stripping of regoliths in conjunction with frost shattering processes within a periglacial environment seems to have been widespread within the British Isles and that the final phase of tor development was intimately related to these conditions.

Some major patterned ground features

Fossil frost wedges

Features resulting from thermally induced contraction processes have been noted in the field for at least a century although their genetic significance was not originally appreciated. Naturally the earliest discussions of their significance were limited by the paucity of data relating to modern ice wedges and allied phenomena. For instance, it was originally assumed that all apparent periglacial wedge structures related to former ice wedges but, following the identification of wedges with primary sediment infills in Antarctica, Worsley (1967) suggested that fossil sand wedges were present at a locality in Cheshire since the texture of the deeper parts of the wedge was inconsistent with an infill resulting from the melt of ground ice. In a discussion appended to this account, Péwé questioned the possibility of sand wedges forming in an environment assumed to have been as maritime then as now. Subsequent investigations in Siberia have shown that both ice and sand wedges may coexist within the same environment, their location determined by microclimatic differences. Hence there can be no *a priori* reason why a complete range of infill types, with the extreme members being ice and sand wedges, should not be present within the British Isles. Further, assessments of palaeoclimate through studies of fossil Coleoptera substantiate the view that periglacial Britain was not always maritime and consequently that primary sediment infills in thermal contraction cracks are entirely feasible. Most of the wedge structures identified so far seem to be epigenetic and this factor could well be a response to relatively slow sedimentation rates.

Non-sorted polygons and stripes

Over the Chalk of eastern England there are large concentrations of non-sorted polygons and stripes with diameters and repeat intervals in the order of 10 m and 7 m respectively. Rather strangely, they are rare in the chalk areas in the more westerly parts of the country and the localization seems to be determined by the existence of a surficial cover of up to 2·5 m of either sandy or loamy materials which constitute a type of coversand. Examples of these structures were subject to an intensive examination by Watt *et al.* (1966), who concluded that cryostatic pressures generated within former active layers were the primary cause of an interaction between the coversand and the underlying chalk. In seeking a present-day analogue, they suggested that the best match was with the Alaskan spotted tundra polygons.

Fossil thermokarst

Interest in fossil thermokarst features was generated by Pissart's (1963) pioneer work in east central Wales. These landforms usually comprise circular or elongate basins of poor drainage surrounded by a rim or rampart whose completeness varies with the general site characteristics. Within some basins up to at least 10 m of organic infill have been encountered. Following Pissart's lead, other possible groupings have been identified in several locations, but so far the richest area has been Wales, although the suspicion arises that the known distribution partly reflects the interests of the particular field workers in each area. In Wales, Watson (1971) has argued that the examples which he has examined in detail are all former open system pingos and convincing parallels have been drawn from central Alaska. In considering rather similar forms in the western parts of East Anglia, Sparks *et al.* (1972) adopted a more cautious approach, prefering to use the non-committal term 'ground ice depressions'. At least with reference to their intensively studied examples this seems to be rather unduly pessimistic, for a comparison with the Alaskan evidence reveals some striking parallels and it is here suggested that at least the fresh East ˙Anglian features are in the main directly comparable with the Welsh fossil pingos. One point of significance may be that Sparks *et al.* orientate their discussion to a consideration of the restricted size of their hollows in comparison with the closed-system pingos as described in the literature. It is doubtful if any open-system pingo remnants will be discovered since there seems to have been a lack of the appropriate kind of low-lying alluvial environment (as exemplified by the modern Mackenzie delta area). Subsurface structures at Wretton in Norfolk have been interpreted as former ground ice mounds but these are of restricted size, being only a few metres in diameter, and cannot be associated with permafrost with any certainty.

Involutions

Under this heading various non-sorted structures, usually observed in vertical section only, will be discussed briefly for they are undoubtedly the commonest form of supposed periglacial structure to be encountered in the field. They are commonly termed involutions by local workers although the term 'festoon' is also used. Usually involutions possess one or more of the following characteristics: (a) a wide range of fold styles extending from gentle undulations to sharp disharmonic piercement structures, (b) restriction to a layer no more than 3 m in depth within which the degree of disruption often decreases with depth, (c) lithologies of contrasting textural properties, and (d) reorientation of the macrofabric such that elongate particles become erect.

It may be argued that not all involutions are indices of periglacial processes for they seem to be uncommon in current periglacial regions. Despite this, the field evidence frequently indicates a strongly positive relationship between involutions and former severe climatic conditions as established on other criteria. They are frequently found at horizons coinciding with the tops of fossil frost wedge structures, suggesting a genetic connection. Hence there is a strong possibility that either active layers or seasonally frozen layers influence their formation and consequently the allied processes of cryostatic pressures and differential frost heave, together with the creation of reversed density stratification, all probably contribute to the final forms. Involutions are surface or near-surface deformations and sub-surface occurrences always reveal evidence for post-formational burial. A comparison of their amplitude with thaw depths in current periglacial environments led Williams (1975) to suggest that they may indicate palaeoclimates with summer temperature regimes which are diagnostic of the continuous/discontinuous permafrost transition.

Fluvial processes

Fluvial activity is a vital component of contemporary periglacial environments, yet the association is often overlooked. This generalization is well exemplified by the limited amount of attention devoted to fluvial processes in the most recent British periglacial synthesis (Embleton and King 1975). Within the former British periglacial zones there can be little doubt that the dominant element of the landscape was the fluvial erosional–depositional system. This factor has been recognized by some workers who have been able to trace the head mantle on slopes into aggradations occupying the valley axes. Thompson and Worsley (1967) interpreted an important Devensian sequence in terms of a prolonged period of fluvial activity contemporaneous with periglacial climates.

Many stratigraphic studies of fossiliferous river terrace sequences, in conjunction with radiocarbon dating, have revealed the considerable extent of fluvial aggradations during the Devensian, especially beyond the direct effects of glaciation *per se*. Sedimentological studies show that for most of the last glacial stage, run-off was re-working the head on valley slopes and transporting it on to wide alluvial tracts across which rivers migrated to and fro. Virtually every Devensian 'interstadial' site records a brief period of stability within part of a river floodplain environment. This pattern of fluvial activity seems to have been common to each of the glacial stages, for when the river systems underwent a metamorphosis, their channel patterns adjusted to both a greatly increased rate of sediment supply and a higher peaked discharged.

By way of illustration we may take the catchment of the River Kennet, a tributary to the River Thames in south central England. This valley was certainly not glaciated in the Upper Pleistocene and possibly not at all, and its position inland has isolated it from the effects of eustatic sea-level change. The present single-channel meandering river is essentially a solute load stream and the Flandrian floodplain deposits are on the whole fine-grained clastics and organic materials. In contrast, the river terraces consist of coarse materials, and the overall geometry of the sedimentary bodies indicates that the terraces represent former low sinuosity palaeoenvironments. On the most recently abandoned aggradational surface a braided palaeo-channel system is still visible (Cheetham 1976). In addition boulder-sized clasts (sarsens) are found throughout the gravels and may well signify that ice rafting was operative concurrently with deposition. During the Devensian, floodplain sediments of the previous inter-glacial were eroded (no representatives have yet been identified) and an extensive but slow aggradation episode followed. In the higher reaches of the catchment, head sheets carrying blockstream covers fed into the bottoms of the valleys. The Flandrian was immediately preceded by a limited amount of incision creating a shallow trench within which the postglacial fluvial sediments lie.

Stratigraphical positions of periglacial phenomena

The important phases of periglaciation are registered by fossil perma-frost structures. Any attempt to assign age limits is clearly dependent upon the precision to which the various stratigraphical components at a given locality can be dated and this in turn requires the most complete successions possible. The East Anglian region naturally pro-vides the best evidence, for there most of the type stage successions are defined. This also has the advantage that if the former presence of permafrost can be established in the south-east of the British Isles, then it may reasonably be supposed that it also affected the northern areas. The converse need not apply and Williams (1969) has proposed a possible east–west contrast in palaeoclimate characteristics which could have played an important role in determining different geo-morphological responses with longitude. However, the validity of this idea is questionable.

Within East Anglia, fossil frost wedges and associated involutions are known from sequences currently identified as representatives of the last five glacial stages, extending down into the Middle Pleisto-cene. Under favourable conditions, the shore of part of the Norfolk coast will expose fossil frost wedge polygons of pre-Cromerian age and sometimes these can be traced and matched with structures seen in vertical section in the cliffs. Owing to the current controversial status of the East Anglian Wolstonian sequence, undisputed evidence for permafrost in the penultimate glacial stage has to be sought in the type area in the English Midlands. There a fossil frost wedge is firmly fixed stratigraphically and a well-developed system within the adjacent Baginton Gravels may be nearly synchronous with the former structure (Shotton 1968).

Devensian periglacial phenomena are the ones most frequently encountered in the field, since evidence for these is normally found at or immediately below the modern landsurface. Aerial photography taken under certain optimal conditions may reveal a patterned ground mosaic over much of the area south of the main glacial limit and also for some distance north of it. Surface developments of permafrost structures can only be related to the last glacial stage south of the glacial limit on geomorphological criteria, but within the limit a greater degree of precision is possible since the structures can be related to the ice wastage chronology. Where exposures of the four most important Devensian marker horizons (i.e. Chelford, Upton Warren *sensu stricto*, and Allerød Interstadials and the till sheet) are available, it is possible to distinguish several phases of permafrost presence both inside and outside the maximal ice advance limit. There remains some doubt as to whether any true pre-Chelford interstadial permafrost structures have as yet been identified but, assuming that the interstadial is represented by one of the organic beds at Wretton, Norfolk, then at least one phase of ice wedge and ground ice mound growth pre-dates it (West *et al.* 1974). At Four Ashes in Staffordshire there is good evidence for a Mid-Devensian phase of frost wedge development (Morgan 1973). Further north, well inside the glacial limit in Cheshire, wedges are associated with an early and Mid-Devensian alluvial fan sequence, but distinctive marker beds are of limited extent and currently it is only possible to assign the structures to a period immediately after the Chelford interstadial and later (Worsley 1966). An extensive network of frost wedge polygons penetrates the Late Devensian till sheet near its limit at Four Ashes and the same relationship applies in Cheshire also. In the latter area the glaciogenic sediments are thicker and intraformational wedges can be recognized showing that the margins of the former glacier probably possessed sub-polar thermal characteristics.

The last, and in many ways the most dramatic, phase of periglaciation was related to the late glacial climatic deterioration which culminated in Zone III at the end of the Devensian. The extent of permafrost growth at this time is still uncertain, for the details are only just emerging, but its presence is now confidently established in Scotland both beyond and inside the glacial readvance limit (Sissons 1976). In northern Ireland a single generation of wedges post-dates the deglacierization and Colhoun (1971) favoured their dating to Zone III. With the renewed development of permafrost in the north the possibility arises that surface outcrops of frost wedges further south may have been reactivated at this time. Support for this possibility may be derived from the findings of Mitchell (1973). Numerous open-system pingos in south-east Ireland were related by him to

Zone III as a result of studies on the pollen in associated sediments. Such facts suggest that there was no major east–west climatic distinction and that severe climatic conditions affected the whole region.

A widespread, sedimentary response to the last major climatic deterioration is that of a palaeosol of Zone I/II age, buried or truncated by higher energy facies. It was first noted in the south east where fans of crudely stratified chalky debris emerged from the coombes cut into the chalk. This pattern has recently been described from the head of the River Gipping estuaries in East Anglia, where a major incision event in Zone III was rapidly followed by a massive aggradation which is thought to indicate a sudden switch to conditions of high sediment yield within the catchment (Wymer *et al.* 1975). Along the Welsh borderland (Rowlands and Shotton 1971) and in the Isle of Man (Mitchell 1965) large amounts of frost-shattered debris were swept out of upland areas on to alluvial fan surfaces. Thus in several areas there appears to have been a significant reactivation of periglacial mass wasting operating in conjunction with a transportation system which is perhaps best described as niveo-fluviatile. The effect of this on the total drainage net is unknown but it was probably not uniform and was controlled by the individual threshold state pertaining within each catchment.

Flandrian and contemporary activity

Active or recently active surface periglacial forms have been reported from most of the higher mountainous tracts. A comprehensive inventory for the north Wales uplands has been presented by Ball and Goodier (1970) and lists compiled in the other regions are similar. Generally a mild periglacial regime now prevails at altitudes in excess of 600 m, although it is stressed that such a figure is not applicable in every case. Permafrost is absent, for even on the coldest summits it is unlikely that the mean annual temperature falls below 9°C.

Sissons has aptly summarized the situation: 'The periglacial landscape of the Scottish mountains is essentially a fossil landscape, modified only in detail (though often intricately) by later periglacial action.' This opinion is largely determined by the striking distribution of large-scale periglacial forms which are outside the limit of the Zone III glacial readvance. Current evidence suggests that during the Flandrian there have been only minor changes in the degree of periglacial activity in comparison with the present. Occasionally man-made structures have been involved in movement and Ball and Goodier were also able to demonstrate that a stone-banked soli-fluction lobe included material derived from a wall constructed prior to A.D. 1815. They, in common with others, have suggested that during the historically recorded period of climatic deterioration, A.D. 1550-1750, active periglacial processes were more extensive than at present. There is scope for the application of lichenometrical techniques in assessing the increase in activity, but no rigorous work has yet been attempted. Some success has been achieved by the radiocarbon dating of fossil soils buried beneath solifluction lobes, and in Scotland mass movement has continued at least intermittently for some 6000 calendar years.

Conclusion

Despite some devotees who emphasize the amount of current periglaciation, there is no way of escaping the fact that approximately 10 000 BP truly periglacial environments within the British Isles were abruptly terminated and have not since returned. Testimony to these former episodes has to be sought through the examination of fossil structures and facies analysis. They can only be satisfactorily interpreted by a comparison with areas outside the British Isles where periglacial climates now prevail. There still remains the difficulty of allowing for a higher-angle sun commensurate with a latitudinal position of some 52°-53°N. Thus it is inevitable that interpretations should always be made with an awareness that the 'key to the past may not always be the present'.

References

Ball, D.F. and Goodier, R. (1970). Morphology and distribution of features resulting from frost-action in Snowdonia. *Fld. Stud.* 3, 193-218.

Chandler, R.J. (1972). Periglacial mudslides in Vestspitsbergen and their bearing on the origin of fossil 'solifluction' shears in low angled clay slopes. *Q. J. Eng. Geol.* 5, 223-41.

Cheetham, G.H. (1976). Palaeohydrological investigations of river terrace gravels. In *Geoarchaeology: Earth science and the past.* (ed. D.A. Davidson and M. Shackley), pp. 335-43. Duckworth, London.

Colhoun, E.A. (1971). Late Weichselian periglacial phenomena of the Sperrin Mountains, northern Ireland. *Proc. R. Ir. Acad.* 71B, 53-71.

Doornkamp, J.C. (1974). Tropical weathering and the ultra-microscopic characteristics of regolith quartz on Dartmoor. *Geogr. Annlr.* 56A, 73-82.

Eden, M.J. and Green, C.P. (1971). Some aspects of granite weathering and tor formation on Dartmoor, England. *Geogr. Annlr.* 53A, 92-9.

Embleton, C. and King, C.A.M. (1975). *Periglacial geomorphology.* Edward Arnold, London.

Fitzpatrick, E.A. (1956). Progress report on the observations of periglacial phenomena in the British Isles. *Biul. peryglac.* 4, 99-115.

Hutchinson, J.N. (1974). Periglacial solifluction: an approximate mechanism for clayey soils. *Géotechnique* 24, 438-43.

Kellaway, G.A. (1973). Development of non-diastrophic Pleistocene structures in relation to climate and physical relief in Britain. *Int. geol. Congr.* 24, (12), 136-46.

Periglaciation

Linton, D.L. (1955). The problem of tors. *Geogrl. J.* 121, 470-87.

Mitchell, G.F. (1973). Fossil pingos in Camaross Townland, Co. Wexford. *Proc. R. Ir. Acad.* 73B, 269-82.

Mitchell, G.F. (1965). The Quaternary deposits of the Ballaugh and Kirkmichael districts Isle of Man. *Q. Jl. geol. Soc. Lond.* 121, 359-81.

Morgan, A.V. (1973). The Pleistocene geology of the area north and west of Wolverhampton, Staffordshire, England. *Phil. Trans. R. Soc., Ser. B,* 265, 233-97.

Mottershead, D.N. (1971) Coastal head deposits between Start Point and Hope Cave, Devon. *Fld. Stud.* 3, 433-53.

Pissart, A. (1963). Les traces de 'pingos' du Pays de Galles (Grande-Bretagne) et du Plateau des Hautes Fagnes (Belgique). *Z. Geomorph.* (N.S.), 7, 147-65.

Rowlands, P.H. and Shotton, F.W. (1971). Pleistocene deposits of Church Stretton (Shropshire) and its neighbourhood. *Q. Jl. geol. Soc. Lond.* 127, 599-622.

Shotton, F.W. (1968). The Pleistocene succession around Brandon, Warwickshire. *Phil. Trans. R. Soc. Ser. B,* 254, 387-400.

Sissons, J.B. (1976). *Scotland.* Methuen, London.

Small, R.J., Clark, M.J. and Lewin, J. (1970). The periglacial rock-stream at Clatford Bottom, Marlborough Downs, Wiltshire. *Proc. Geol. Ass.* 81, 87-98.

Sparks, B.W., Williams, R.G.B. and Bell, F.G. (1972). Presumed ground ice depressions in East Anglia. *Proc. R. Soc. Ser. A,* 327, 329-43.

Thompson, D.B. and Worsley, P. (1967). Periods of ventifact formation in the Permo-Triassic and Quaternary of the north-east Cheshire Basin. *Mercian. Geol.* 2, 279-98.

Washburn, A.L. (1973). *Periglacial processes and environments.* Edward Arnold, London.

Waters, R.S. (1964). Great Britain. *Biul. peryglac.* 14, 109-10.

Watson, E. (1966). Two nivation cirques near Aberystwyth, Wales. *Biul. peryglac.* 15, 79-98.

Watson, E. (1971). Remains of pingos in Wales and the Isle of Man. *Geol. J.* 7, 381-92.

Watson, E. and Watson, S. (1967). The periglacial origin of the drifts at Morfa-Bychan, near Aberystwyth. *Geol. J.* 5, 419-440.

Watt, A.S., Perrin, R.M.S. and West, R.G. (1966). Patterned ground in Breckland: structure and composition. *J. Ecol.* 54, 239-58.

West, R.G., Dickson, C.A., Catt, J.A., Weir, A.H. and Sparks, B.W. (1974). Late Pleistocene deposits at Wretton, Norfolk. *Phil. Trans. R. Soc., Ser. B,* 267, 337-420.

Williams, R.G.B. (1968). Some estimates of periglacial erosion in southern and eastern England. *Biul. Peryglac.* 17, 311-35.

Williams, R.G.B. (1969). Permafrost and temperature conditions in England during the last glacial period. In *The periglacial environment* (ed. T.L. Péwé), pp. 339-410. Mc-Gill-Queens University Press, Montreal.

Williams, R.G.B. (1975). The British climate during the Last Glaciation; an interpretation based on periglacial phenomena. *Geol. J. Spec. Issue* 6, 95-120.

Worsley, P. (1966). Some Weichselian fossil frost wedges from East Cheshire. *Mercian Geol.* 1, 357-65.

Worsley, P. (1967). Fossil frost wedge polygons at Congleton, Cheshire, England. *Geogr. Annlr.* **48A**, 211-9.

Wymer, J.J., Jacobi, R.M. and Rose, J. (1975). Late Devensian and Early Flandrian barbed points from Sproughton, Suffolk. *Proc. prehist. Soc.* **41**, 235-41.

16

Loess and coversands

J. A. CATT

Abstract

Thin loess, deposited during the early part of the Late Devensian, and often mixed by cryoturbation with subjacent deposits, is widespread in parts of south and east England and south Wales outside the Devensian glacial limit. Originally more extensive, it has been removed from many areas by Late Devensian solifluction, Late Devensian and early Flandrian stream erosion, and colluviation resulting from Neolithic and later deforestation and agriculture. Reasons for its preferred stability on limestone surfaces are discussed. Isolated deposits of older (pre-Ipswichian) loess also occur in the Thames and Gipping valleys and on the Chiltern Hills.

Most of the coversands in Lincolnshire, Yorkshire, and Lancashire are equivalent in age to Younger Coversand II of The Netherlands, but there are older Coversands in East Anglia.

Introduction

Periglacial aeolian deposits are quite widespread in parts of England and Wales (Fig. 16.1), but have only recently been recognized and described, because they are often thin, weathered, and reworked. Prestwich (1864) noted the similarity of many of the river deposits in southern England to the German loess, and isolated loess deposits have long been known in Co. Durham (Trechmann 1920), Hampshire

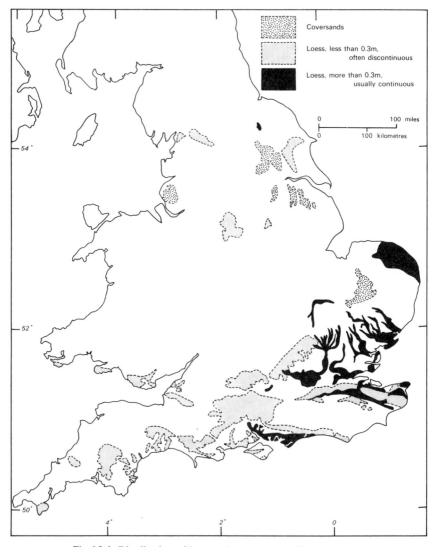

Fig. 16.1. Distribution of loess and coversands in England and Wales.

(Palmer and Cooke 1923) and Kent (Pitcher *et al.* 1954). However, the widespread occurrence of thin, weathered loess has only been recognized more recently through pedological work, mainly by the Soil Survey of England and Wales. Similarly, Pleistocene aeolian sands were described in Somerset by Greenly (1922) and Lincolnshire by Straw (1963), but recent soil mapping and profile studies have shown that they also cover large areas of East Anglia, Yorkshire and Lancashire.

Because many of the aeolian deposits are thin, weathered throughout, mixed with other materials by cryoturbation or mass movement, or have been re-deposited by water, few fit the fairly strict definitions of loess and coversand suggested by Russell (1944), Van der Hammen (1951), and others. However, their ultimate aeolian origin is indicated sometimes by mineralogical and granulometric affinities with deposits that do fit the stricter definitions, and sometimes by field relationships indicating derivation from sediment that occurs in so many different physiographic situations that it could only have been deposited by wind.

Distribution of loess and coversands

Unweathered calcareous loess, which has not been reworked by fluvial or colluvial processes, forms a continuous deposit over Chalk, Lower Tertiary sediments, and river terrace gravels in areas adjacent to the Thames estuary (north Kent and south-east Essex), but occurs only rarely elsewhere. Geologists have mapped and described this as 'brickearth' or 'head brickearth' (Dines *et al.* 1954; Smart *et al.* 1966; Gruhn *et al.* 1974). The soils in it, and in deposits >1 m thick thought to be composed mainly of reworked loess, are mapped by the Soil Survey as Hamble, Hook, or Park Gate series, depending on the extent of gleying in the profile (Hodgson 1967). In soils with thinner (0·3–1·0 m) loess-derived horizons, the profile features are determined largely by the nature of the substratum, and many more series are recognized (Catt 1977). These occur on the Carboniferous Limestone in Somerset, South Wales, Derbyshire, Yorkshire, and Westmorland, on Clay-with-flints and related superficial deposits covering Chalk plateau surfaces in southern England, on chalky solifluction deposits, and on many Chalk surfaces devoid of other superficial materials. West of the Chalk outcrop in Dorset, where the Clay-with-flints-and-cherts extends on to the Upper Greensand on the east Devon plateau, it is mantled by thin loess, and a similar deposit occurs on the Budleigh Salterton Pebble Beds and Haldon Gravels in east Devon, on the Devonian Limestones near Torquay, and in solifluction (head) deposits on the Dartmoor Granite (Harrod *et al.* 1973).

Mottershead (1971) described loess horizons in the coastal head deposits of south Devon; a thin layer also overlies the Cornish serpentine (Coombe *et al.* 1956) and head deposits on the Welsh coast near Aberystwyth (Watson and Watson 1967). Loess on the Tertiary deposits in south Hampshire was mapped as 'brickearth' by the Institute of Geological Sciences and by Fisher (1971). Catt *et al.* (1971) recognized widespread loess in soils over older glacial deposits in north-east Norfolk; a similar 'coverloam' occurs in north-east Essex, but in other parts of East Anglia loess occurs mainly in the river valleys. Thin isolated deposits also occur on the Bunter Sandstone in Nottinghamshire, and on Corallian Limestones of the Hambleton Hills, north Yorkshire (Catt *et al.* 1974).

The aeolian sands in Lincolnshire described by Straw (1963) are part of an extensive sheet of variable thickness occupying the lower Trent valley and southern parts of the Vale of York (Matthews 1970, 1971; Gaunt *et al.* 1971); locally they form dunes up to 6 m high. The Shirdley Hill Sands of south-west Lancashire are also probably aeolian, though dunes are rare (Hall and Folland, 1967). In the Breckland area of East Anglia the Chalk is covered by a widespread chalk-sand drift, which was originally interpreted as a local facies of the Gipping Till (Baden-Powell 1948; Perrin 1956), but more recently as a periglacial deposit composed of cryoturbated and soliflucted chalk and coversand (Watt *et al.* 1966; Corbett, 1973).

By analysing topsoil samples taken from one or more interfluve sites per 100 km² in eastern and southern England, Perrin *et al.* (1974) linked Yorkshire, Lincolnshire, and the Breckland into one coversand province, and showed that this is bounded to the north-east and south-west by provinces in which the aeolian deposits are silt–sand mixtures. From their work, the only areas with silt and no aeolian sand are parts of the Yorkshire and Lincolnshire Wolds and the whole of southern England south of a line approximating to the Thames valley.

Stratigraphy and dating

Almost all the loess in Britain is attributable to the earlier part of the Late Devensian, but only on indirect evidence, as there are few stratigraphically significant features. In Kent, solifluction deposits containing loess underlie pellety chalk muds of the Older Dryas period (Kerney 1963, 1965), and in Yorkshire similar deposits lie between the Late Devensian till and the Sewerby (Ipswichian) interglacial sediments (Catt *et al.* 1974). As the mineralogical composition of the coarse silt in the loess is similar, especially in eastern England, to the same size fraction of Late Devensian glacial deposits, most of the loess was probably derived from glacial outwash in the North Sea basin, and is therefore approximately contemporaneous with that ice advance. The modal size of the silt decreases westwards across southern England from 45 to 25 μm, and there is an increase in the same direction of flaky silt minerals (micas and chlorite); both changes suggest that the loess was transported westwards across the country from the North Sea source area by dominantly easterly winds (Catt 1977).

A Late Devensian age is also indicated by comparison with adjacent areas of continental Europe, where the only widely distributed loess is Upper Pleniglacial, by the frequent incorporation of the loess in solifluction deposits and periglacial soil structures, such as involutions and frost wedges, and by archaeological evidence. In Hampshire the 'brickearth' and associated chalky solifluction deposits are younger than Aurignacian (Palmer and Cooke 1923), and at several cave sites in south-west England, south Wales, and the Midlands, loess-containing deposits overlie Upper Palaeolithic industries.

However, isolated deposits of an older (pre-Devensian) loess occur at several sites in south-east England. At Northfleet (Kent) there are two 'brickearths', the older of which contains Mousterian implements (Burchell 1933) and *Pupilla muscorum* (L.), *Vallonia costata* (Müll.), and *Succinea oblonga* (Drap.) (Burchell 1954), yet underlies a fluviatile mud with an Ipswichian mollusc fauna. Loess also occurred beneath Ipswichian interglacial deposits at Bobbitshole, Ipswich (West 1958), and a similar fine calcareous silt underlies Chalky Boulder Clay in the Gipping valley, Suffolk (Rose and Allen 1977). On the Chiltern Hills, silty deposits called 'true brickearth' by Barrow (1919) fill doline-like depressions in the Clay-with-flints, and locally contain Acheulian industries (Smith 1916). The 'interglacial' loess of Durham (Trechmann 1920) lies between two tills, which Francis (1970) correlated with the Wolstonian and the Late Devensian glacial stages. He suggested that the loess is Wolstonian, as erratic boulders in its upper layers were weathered in a warm (interglacial?) period, but it could also be Late Devensian if the erratics were incorporated during that glaciation.

Most of the English coversands seem to be younger than the loess. In Yorkshire and Lancashire they are above rather than beneath the Late Devensian till, and radiocarbon dates of interbedded peats in Yorkshire (Gaunt *et al.* 1971) and Lincolnshire (Shotton and Williams 1973) suggest that this aeolian phase was after 11 000 BP. Most of the sands are therefore correlated with Younger Coversand II of The Netherlands, but subsequent reworking has occurred widely, and even continues today during dry, windy periods. Perrin *et al.* (1974)

disagree with the relative ages of the loess and coversand stated above; they believe the coversands are older, because in East Anglia they are involved more frequently in involutions, and are locally overlain by silt, and they suggest the silt-sand deposits result from biological mixing of loess laid on pre-existing coversand. However, the loess in many other areas is periglacially disturbed, and the superposition of loess on coversand may be because at least part of the coversand in East Anglia is Early Devensian (West *et al.* 1974), and therefore equivalent to the Older Coversands of The Netherlands.

Pleistocene ventifact horizons occur in Lincolnshire, Nottinghamshire, Worcestershire, Lancashire, Cheshire, and in many parts of the Vale of York. Like the coversands, they seem to be locally of two ages, one older and one younger than the Late Devensian glacial phase (Gaunt 1970).

Flandrian erosion and soil formation

Although loess is more widespread in England and Wales than was previously thought, it is less continuous than an aeolian deposit of Late Devensian age should be, and must have suffered extensive erosion during Late Devensian and Flandrian times. Assuming that a fairly even blanket of loess was deposited on all rock types in eastern and southern Britain during the Late Devensian, there has been preferential removal of loess from clayey and sandy substrata. This is especially evident in the Midlands and parts of northern England, where the loess is restricted almost without exception to non-dolomitic limestone surfaces. Some erosion of loess undoubtedly occurred in periglacial conditions during the Late Devensian, as it is a common constituent of solifluction deposits in many areas. Further downslope movement resulted from soil degradation by Neolithic and later agriculture; some of this was left as footslope colluvium, and some was redeposited as alluvium. However, it is unlikely that solifluction and agriculturally induced soil erosion explain the absence of loess from large areas, such as parts of the Weald, the Midland clay vales, the Millstone Grit, Coal Measure and Magnesian Limestone outcrops of the north Midlands and Yorkshire, and the Culm and Devonian outcrops of north Devon, which are all south of the Late Devensian ice limit and must have received some loess at that time.

The only way in which loess could have been stabilized on limestone surfaces while it was eroded almost completely from other substrata is by secondary carbonate cementation of subsoil horizons in periglacial deposits composed of loess and frost-shattered limestone. Such cementation is most likely to have occurred early in the history of soil development, and this probably limits it to the last few thousand years of the Devensian or the first few thousand of the Flandrian, because in east Kent decalcification of a loess soil and much subsequent clay eluviation had occurred before the Atlantic period (Weir *et al.* 1971). Also, widespread erosion could only have occurred before the soils were stabilized by the main Flandrian forest development. Little or no cementation would have occurred in loess itself, or in head deposits formed on non-calcareous rocks; periglacial deposits on the Magnesian Limestone would also have remained uncemented, because dolomite is much less soluble than calcite in temperate soils.

The removal of loess above clay substrata probably resulted mainly from stream erosion, to which it is especially susceptible. Although it is very porous and has a low bulk density, loess is extremely firm when dry or slightly moist, probably because the silt particles are propped apart by clay bridges, the orientation of which may have been determined by frost during deposition. However, flooding rapidly breaks down this structure, and disaggregates the loess almost to its primary constituent particles, which are easily transported by water. This erosion would have been rarer on more permeable substrata such as sandstones, though some sheet erosion of surface soil in such areas might have resulted from the slight drainage impedance that can occur when clay is eluviated from higher soil horizons and redeposited in subsoil drainage channels. The friable, clay-depleted surface horizons of some soils may also have been subject to wind erosion in drier periods. The Flandrian alluvial deposits of many English rivers are very silty, and probably contain some of the loess that was eroded by streams and surface run-off, though most was probably carried into the sea.

Conclusions

There is much to be learnt about the origin, conditions of deposition, weathering, erosion, and redistribution of aeolian deposits by studying them where they are thin and rather discontinuous, as in Britain. They are so, partly because Britain was at the margin of the European loess and coversand belt and the supply of aeolian sediment was limited, but also because considerable erosion occurred during and after deposition, largely as a result of Britain's fairly oceanic climate. The history of erosion, weathering, and redeposition is evidently complicated by variations in bedrock geology, by changes in past climate and natural vegetation, and by chapters of human activity. However, all these factors can be evaluated independently, because the aeolian deposits are only a minor part of the British Quaternary sequence. British workers are beginning to study their importance in determining the present nature and local distribution of loess and coversand, and it is hoped that in time this will help explain some of the puzzling features of much thicker sequences abroad.

References

Baden-Powell, D.F.W. (1948). The chalky boulder clays of Norfolk and Suffolk. *Geol. Mag.* 85, 279-96.

Barrow, G. (1919). Some future work for the Geologists' Association. *Proc. Geol. Ass.* 30, 1-48.

Burchell, J.P.T. (1933). The Northfleet 50-foot submergence later than the Coombe Rock of post-Early Mousterian times. *Archaeologia* 83, 67-92.

Burchell, J.P.T. (1954). Loessic deposits in the Fifty-foot Terrace post-dating the Main Coombe Rock of Baker's Hole, Northfleet, Kent. *Proc. Geol. Ass.* 65, 256-61.

Catt, J.A. (1977). The contribution of loess to soils in lowland Britain. *C.B.A. Res. Rept.*

Catt, J.A., Corbett, W.M., Hodge, C.A.H., Madgett, P.A., Tatler, W. and Weir, A.H. (1971). Loess in the soils of north Norfolk. *J. Soil Sci.* 22, 444-52.

Catt, J.A., Weir, A.H. and Madgett, P.A. (1974). The loess of eastern Yorkshire and Lincolnshire. *Proc. Yorks. geol. Soc.* 40, 23-39.

Coombe, D.E., Frost, L.C., Le Bas, M. and Watters, W. (1956). The nature and origin of the soils over the Cornish serpentine. *J. Ecol.* 44, 605-15.

Corbett, W.M. (1973). Breckland Forest Soils. *Soil Surv. Spec. Surv.* 7.

Dines, H.G., Holmes, S.C.A. and Robbie, J.A. (1954). Geology of the country around Chatham. *Mem. geol. Surv. U.K.*

Fisher, G.C. (1971). Brickearth, and its influence on the character of soils, in the south-east New Forest. *Pap. Proc. Hampsh. Fld. Club* 28, 99-109.

Francis, E.A. (1970). Quaternary. In: Geology of County Durham. *Trans. nat. Hist. Soc. Northumb.* 41, 134-52.

Gaunt, G.D. (1970). An occurrence of Pleistocene ventifacts at Aldborough near Boroughbridge, West Yorkshire. *J. Earth Sci., Leeds,* 8, 159-61.

Gaunt, G.D., Jarvis, R.A. and Matthews, B. (1971). The late Weichselian sequence in the Vale of York. *Proc. Yorks. geol. Soc.* 38, 281-84.

Greenly, E. (1922). An aeolian Pleistocene deposit at Clevedon. *Geol. Mag.* 59, 365-76, 414-21.

Gruhn, R., Bryan, A.L. and Moss, H.A. (1974). A contribution to Pleistocene chronology in south-east England. *Quaternary Res.* 4, 53-71.

Hall, B.R. and Folland, C.J. (1967). Soils of the south-west Lancashire coastal plain. *Mem. Soil Surv. U.K.*

Harrod, T.M., Catt, J.A. and Weir, A.H., (1973). Loess in Devon. *Proc. Ussher Soc.* 2, 554-64.

Hodgson, J.M. (1967). Soils of the West Sussex Coastal Plain. *Bull. Soil Surv. G.B. (Engl. and Wales),* 3.

Kerney, M.P. (1963). Late glacial deposits on the Chalk of south-east England. *Phil. Trans. R. Soc.* B 246, 203-54.

Kerney, M.P. (1965). Weichselian deposits in the Isle of Thanet, East Kent. *Proc. Geol. Ass.* 76, 269-74.

Matthews, B. (1970). Age and origin of aeolian sand in the Vale of York. *Nature, Lond.* 227, 1234-36.

Matthews, B. (1971). Soils in Yorkshire I, Sheet SE65 (York East). *Soil Surv. Rec.* 6.

Mottershead, D. (1971). Coastal head deposits between Start Point and Hope Cove, Devon. *Fld. Stud.* 3, 433-53.

Palmer, L.S. and Cooke, J.H. (1923). The Pleistocene deposits of the Portsmouth district and their relation to man. *Proc. Geol. Ass.* 34, 253-82.

Perrin, R.M.S. (1956). The nature of 'Chalk Heath' soils. *Nature, Lond.* 178, 31-2.

Perrin, R.M.S., Davies, H. and Fysh, M.D. (1974). Distribution of late Pleistocene aeolian deposits in eastern and southern England. *Nature, Lond.* 248, 320-24.

Pitcher, W.S., Shearman, D.J. and Pugh, D.C. (1954). The loess of Pegwell Bay and its associated frost soils. *Geol. Mag.* 91, 308-14.

Prestwich, J. (1864). On the loess of the valleys of the south of England, and of the Somme and the Seine. *Phil. Trans. R. Soc.* 154, 247-309.

Rose, J. and Allen, P. (1977). Middle Pleistocene stratigraphy in south-east Suffolk. *J. geol. Soc. Lond.* 133, 83-102.

Russell, R.J. (1944). Lower Mississippi valley loess. *Bull. geol. Soc. Am.* 55, 1-40.

Shotton, F.W. and Williams, R.E.G. (1973). Birmingham University Radiocarbon Dates VII. *Radiocarbon* 15, 458.

Smart, J.G.O., Bisson, C. and Worssam, B.C. (1966). Geology of the country around Canterbury and Folkestone. *Mem. Geol. Surv. U.K.*

Smith, W.G. (1916). Notes on the Palaeolithic floor near Caddington. *Archaeologia* 67, 49-74.

Straw, A. (1963). Some observations on the Cover Sands of north Lincolnshire. *Trans. Lincs. Nat. Un.* 15, 260-69.

Trechmann, C.T. (1920). On a deposit of interglacial loess, and some transported preglacial freshwater clays on the Durham coast. *Q. J. Geol. Soc. Lond.* 75, 173-201.

Van der Hammen, T. (1951). Late-glacial flora and periglacial phenomena in the Netherlands. *Leid. geol. Meded.* 17, 71-183.

Watson, E. and Watson, S. (1967). The periglacial origin of the drifts at Morfa Bychan near Aberystwyth. *Geol. J.* 5, 419-40.

Watt, A.S., Perrin, R.M.S. and West, R.G. (1966). Patterned ground in the Breckland: structure and composition. *J. Ecol.* 54, 239-58.

Weir, A.H., Catt, J.A. and Madgett, P.A. (1971). Postglacial soil formation in the loess of Pegwell Bay, Kent (England). *Geoderma* 5, 131-49.

West, R.G. (1958). Interglacial deposits at Bobbitshole, Ipswich. *Phil. Trans. R. Soc.* B 241, 1-31.

West, R.G. Dickson, C.A., Catt, J.A., Weir, A.H. and Sparks, B.W. (1974). Late Pleistocene deposits at Wretton, Norfolk. II. Devensian deposits. *Phil. Trans. R. Soc.* B 267, 337-420.

17

A British ice-sheet model and patterns of glacial erosion and deposition in Britain

G.S. BOULTON, A.S. JONES, K.M. CLAYTON, and M.J. KENNING

Abstract

A steady-state model is constructed of the Late Devensian ice sheet which covered much of Britain. Its summit height is 1800 m, velocities in marginal areas range from 150 to 500 m/year, and the basal ice is cold in the central area and temperate near to the margins. About 15 000 years would have been required for its build-up. It is possible that the ice sheet never reached a steady state; its expansion may have been stopped by a rapid climatic amelioration. The lobe of ice which thrust down the east coast of England is suggested to have been emplaced by a surge.

Patterns of glacial erosion and deposition over Britain are summarized. It is suggested that central parts of the ice sheet were relatively inactive and that the intense glacial erosion in highland Britain was not produced during glacial maxima. Relatively high erosional intensities in marginal areas are thought to have been produced by high marginal velocities. These latter also inhibited thick lodgement till and drumlin formation near to the margin, and these are therefore concentrated in internal zones.

Introduction

The profiles and dynamic and thermal regimes of former ice-sheets can be reconstructed if we know their extent and the length of flow lines and if glacier budgets and palaeotemperatures can be estimated. We here present the results of a modelling exercise for the ice sheet which attained a maximum extent over Britain at about 18 000 BP. These results are useful in helping to understand the relationships between climatic and glacial changes at this time, and inferred glacier dynamic and thermal structures can help to explain some of the observed patterns of glacier erosion and deposition in Britain.

The ice-sheet model and some implications

We have used Weertman's (1961a) analysis to determine the surface shape and velocity distribution of the late Devensian ice-sheet over Britain and our own analysis to determine basal temperatures. We assume that at its maximum it was in a steady state, that is that the margin was stationary, ablation balanced accumulation, the glacier flow field was constant with time, and the ice sheet was in thermal equilibrium with the atmosphere and its bed. The predictions made by the model enable us to assess if these assumptions are realistic. The details of the analysis will appear elsewhere (Boulton and Jones, in preparation).

The input data for this model are the following:

(a) The maximum extent of the Late Devensian ice-sheet is shown in Fig. 17.1. It is largely based on Charlesworth's (1957) synthesis, although we have taken Synge's (1969) view of the ice-sheet margin in Ireland, John's (1970) view of the margin in south-west Wales, and a concensus view of the margin in the southern North Sea (see Chapter 14). Much of the continental shelf margin is speculative, but we assume a zone of confluence with Scandinavian ice in the northern North Sea based on the apparent deflection of ice both to the north and south in east Scotland and north-east England, and Hoppe's (1974) suggestion that ice moved over Shetland in a westerly direction during the last glaciation.

(b) Flow lines have been inferred from erratic boulder distributions, till fabric studies, and the orientation of drumlins and glacial striae (Fig. 17.1.). Such patterns reflect net transport directions over several episodes of glaciation rather than flow lines for one episode, and so we have used as a general constraint that our model should not cause flow in directions in which there is no evidence of erratic movement.

(c) Climatic snowlines during the late Devensian glaciation have been estimated from various sources: the summary of Tricart (1970),

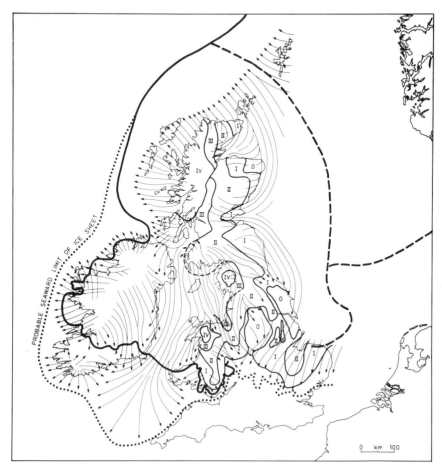

Fig. 17.1. Pattern of ice-sheet movement over Britain and patterns of glacial erosion (excluding Ireland). The dotted line shows the maximum extent of glaciation in Britain. The continuous line shows the assumed limit of the late Devensian ice-sheet, dashed in the speculative North Sea area and omitted over central England. The arrows show generalized flow directions based on geological evidence. Zones of increasing erosional intensity, from 0 to IV are also shown.

the work of Farrington (1953), Manley (1949) and our own observations in south-west Ireland.

(d) The distribution of accumulation has been estimated from the climatic modelling work of Williams and Barry (1974) which yields a prediction of precipitation values over ice-age Britain (estimated as 1800 mm in south-west Ireland decreasing to 700 mm in north-east Scotland) and a lapse rate. From this, and from estimates of July mean temperatures in central England by Williams (1975), Coope

(1975), and Lamb and Woodroffe (1970), we have estimated the distribution of net accumulation.

(e) From these latter estimates it has also been possible to estimate surface temperatures in the ice sheet below the level of annual temperature fluctuations by using modern analogues.

The results of this modelling exercise are shown in Figs. 17.2 and 17.3.

Using the derived climatic parameters, we find that for profiles to the south of the ice-shed terminating in the southern Irish Sea and the English Midlands, values of source height exceed those for the northerly profiles terminating on the Inishowen peninsular in northern Ireland by some 130 m. The most likely explanation of this is

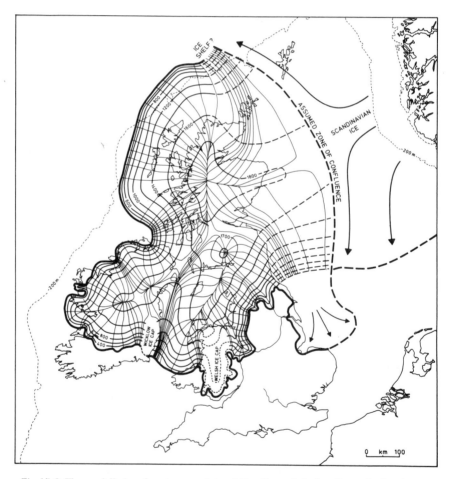

Fig. 17.2. The modelled surface topography and flow lines of the late Devensian ice-sheet.

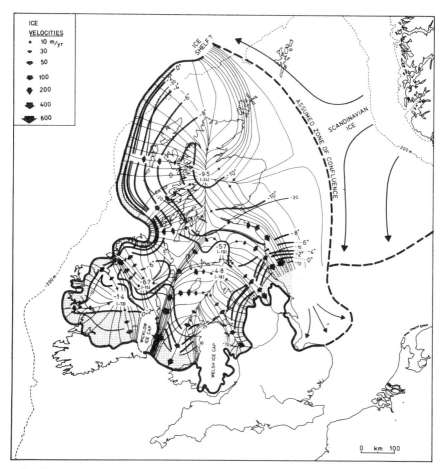

Fig. 17.3. The distribution of mass balance velocities (m/year) for the modelled late Devensian ice-sheet, the basal temperature distribution (°C) (thick lines) and the assumed surface temperature distribution (thin lines and in brackets). The fluctuations of surface temperature near to the margin are too complex to be shown in full. The stippled area represents basal ice at the melting point.

that the geological assessment is incorrect and Inishowen was glaciated during the maximum of the Devensian glaciation. The southern profiles have been adjusted to fit those to the north in Fig. 17.2.

The lobe of ice which is thought to have thrust down the east coast of England cannot be adequately modelled by our method unless we make the climatically implausible assumption that it was a separate centre of ice sheet accumulation. It is more plausibly explained as a surge lobe emplaced as a result of some instability in an ice sheet margin whose steady state position crossed the English coast in the vicinity of the river Tees. Such a possibility has implications for the dating of the maximum extension of Devensian ice in Britain. The lower age limit for this has hitherto been given by the dates of 18 240 ± 250 BP and 18 500 ± 400 BP on a moss bed which underlies a thick till sequence at Dimlington (Humberside) in south-east Yorkshire deposited by the ice of this suggested surge. Such a surge may have occurred before, during or after the period of maximum glaciation, and thus this latter may pre-date, post-date or be synchronous with 18 000 years BP.

In addition to those referred to above, there are several other important features of the model:

(a) Using the precipitation values suggested by Williams and Barry (1974), the model gives a high ablation gradient and a glacier of high activity, with velocities comparable with those of active modern alpine glaciers.

(b) The ablation rate required to maintain a steady state is in places implausibly high; to the west of Scotland in the range 33–42 m/year, in the North Sea area up to 43 m/year, and in the Irish Sea up to 77 m/year. This may be explained in three ways: (i) the precipitation values suggested by Williams and Barry (1974) are much too high; (ii) the high ablation rates might reflect iceberg calving into water in areas where isostatic depression beyond the margin of the ice-sheet was greater than eustatic sea-level lowering. (iii) the glacier never achieved a steady state and might have advanced further had climatic conditions been maintained, but was halted at the observed limit by a rapid climatic amelioration.

Of these three possibilities, we believe the second and third to be most likely.

(c) The total volume of the modelled British ice-sheet, as far east as the assumed zone of confluence with Scandinavian ice, is 346 000 km^3, equivalent to a world-wide lowering of sea-level of 0·96 m. At a mean accumulation rate of 0·1 m/year, over the whole of the ice-sheet area, approximately 15 000 years would be required for its build up.

Rocks eroded directly beneath the postulated ice-dome in the Southern Uplands would take some 19 000 years to reach the ice-sheet margin on the east coast of England, whilst rocks eroded from points 15 km, 50 km, 100 km, and 150 km from this source would take 10 000, 2800, 850 and 187 years respectively to reach the same ice-margin. As 19 000 years is a very substantial part of a glacial period, one would not expect to find rocks eroded beneath the source areas of the ice-sheet to be widely distributed.

(d) The derived basal temperature distribution shown in Fig. 17.3 shows· a clear differentiation between an inner zone in which the basal temperatures are below the melting point, and an outer zone of melting, a pattern which is primarily dependent on surface (~ 10 m) temperature and accumulation rate. The greatest error is likely to arise from errors in the estimated magnitude and pattern of net accumulation which shows an increase from the equilibrium line to the source areas. Normally accumulation rates decrease towards a glacier source, and if the net accumulation rate on the summit dome of the ice-cap over the Scottish Highlands were reduced to 0·24 m/year, the basal temperature would rise to the melting point. Notwithstanding the potential errors, we believe that the marginal pattern suggested here, of an outer zone of melting succeeded up-glacier by a zone below freezing is likely to be substantially correct, although low accumulation rates in source areas may also produce basal melting there.

In our model we assume a steady thermal state. However, we know that many areas of central England were underlain by perma-frost during the long Upton Warren Interstadial which preceded the Late Devensian ice advance. If a glacier melting basally in its marginal zone were to move over an area of permafrost, the effect would be to depress temperatures below the melting point in the extreme mar-ginal zone. Thus, one might expect in general to find the extreme outer basal zone, beyond the zone of basal melting, to be underlain by permafrost with temperatures below the melting point. The effect of this might be, as has been suggested by Weertman (1961*b*) and Boulton (1972*a*), for water produced in the zone of basal melting to be refrozen and re-incorporated in the base of the glacier in the outer zone of freezing.

The geological and geomorphic results of glaciation

In this section we attempt to describe the principal large-scale patterns of erosion and deposition by ice in Britain, and then examine the extent to which they can be related to the proposed ice-sheet model. We shall refer only to Late Devensian patterns of glacial deposition, but as erosional patterns reflect a net landscape response to moulding during a number of successive glacial episodes, we shall assume that the nature of the British ice-sheet was similar to that modelled here during each of the glacial maxima when ice penetrated far south into Britain.

Glacial erosion

Erosion is conveniently mapped by a series of zones (Clayton 1974), which comprise five categories of modification by ice, from Zone 0 (virtually unmodified) to Zone IV, where the entire landscape has

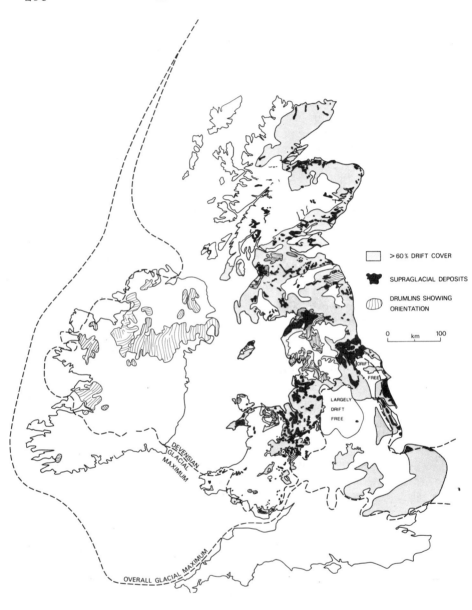

> 60% DRIFT COVER

SUPRAGLACIAL DEPOSITS

DRUMLINS SHOWING ORIENTATION

0   km   100

DRIFT FREE

LARGELY DRIFT FREE

DEVENSIAN GLACIAL MAXIMUM

OVERALL GLACIAL MAXIMUM

Fig. 17.4. The extent of glacial deposits over Britain (excluding Ireland), and the distribution of drumlins (showing mean orientation) and supraglacial sediments (excluding Ireland) within the late Devensian ice-sheet limit. (Some of the data are derived from an unpublished manuscript copy of the Geomorphological Map of Britain kindly lent by Professor E.H. Brown).

been shaped by ice (Fig. 17.1). The criteria are empirical, but the resulting regionalization is a useful index of the extent to which a landscape has been eroded by ice. On the whole the scheme is appreciably easier to apply in upland than in lowland areas.

The most intense erosion has occurred in highland areas, but it will be noted that several of the lowland areas of erosion lie towards the margin of the ice-sheet.

An order of magnitude estimate of the amount of glacier erosion can be obtained from an estimate of the volume of till and associated gravels, which we would estimate at about $2000\,km^3$. In eastern England the volume of the Chalky Boulder Clay alone is of the order of $300\,km^3$, and since about a quarter of this is Chalk, and thus demonstrably of local origin, the ice-sheet must have effected considerable erosional changes on the landforms of the area in this marginal zone of the ice sheet. Indeed, as much of the matrix consists of Jurassic clay (largely the Oxford and Kimmeridge clays), it seems that both the cuestas and the intervening clay vales of the East Midlands and East Anglia must have been greatly modified by the ice-sheet.

The southernmost occurrence of Zone II in England is a fine example of glacial erosion within a lowland landscape. There is a striking change in the chalk escarpment where this crosses the limit of glaciation. To the south the scarp has an altitude of about 250 m. To the north it is broad and irregular and far lower, descending to 90 m in northern East Anglia, suggesting that the crest of the escarpment has been lowered by the eroding ice by 100 m or more. There is evidence that the scarp has been set back by about 3 km. This eroded escarpment zone may be compared with the complementary glacial deposits which cover a belt up to 10–20 km wide on the Chalk dipslope. The eroded and deposited volumes are of the same order, although since only about one-third of the till and gravel is Chalk, it seems that only about half the eroded material is retained on the dip slope; the rest was presumably lost to the North Sea basin as meltwater carried sediment and dissolved carbonate away from the ice-sheet. It is interesting to note that just to the north of the section where the scarp shows signs of glacial erosion, Bromley (1967) described the southward and upward displacement of huge masses of Chalk.

An explanation of observed erosional patterns

It has been suggested (Boulton 1974) that the processes of glacier erosion beneath temperate ice depend primarily on the velocity of ice movement over the bed and the effective pressure at the glacier/

bed interface. Increased ice velocity will increase the horizontal trac-
tive force on subglacial materials and thus favour erosion, whereas
increased effective pressure (ice pressure — water pressure) will increase the frictional resistance to movement of subglacial materials
and thus tend to inhibit erosion. Although velocity fluctuations are
predicted in the model, effective pressures are more difficult to estimate. Weertman (1972) concludes that for an impermeable bed
effective pressures will be constant if bed roughness is constant,
whilst Boulton (1975) suggests that for a permeable bed, effective
pressures at any one point depend principally upon the integrated
hydrogeological characteristics of the bed. If we assume that there is
a random fluctuation in effective pressure in permeable beds, then
we expect that in areas of basal melting broad patterns of erosion are
determined largely by ice velocity. Indeed beneath modern glaciers
we find the bed is eroded and swept clear of any till if velocities are
high, whilst till is deposited upon it if velocities are low.

Our understanding of erosional processes beneath cold ice is much
more limited. It is generally considered that cold ice will not slide
over its bed because of the relatively high adhesive forces between
the two, and thus that many erosional processes will be partly or
completely suppressed, although basal adhesion may lead to tearing
up of bed materials if their cohesion is low and effective pressures
are small. Where cold-based ice occurs down-glacier of a zone of
basal melting, net freezing-on of meltwater will tend to incorporate
loose detritus or unlithified sediments into the basal part of the glacier.

If we attempt to apply these ideas to the map (Fig. 17.1) showing
zones of erosion in the light of the ice-sheet model (Figs. 17.2 and
17.3) it becomes clear that conditions under the centres of ice outflow, the Scottish Highlands, the Lake District, North Wales, and the
Southern Uplands (low ice velocities and possibly sub-melting point
temperatures), would not be conducive to high rates of erosion by
any process, although, except in the case of the Southern Uplands,
these are the areas which show evidence of most intense (Zone IV)
erosion. We would conclude from this that erosion in these areas was
not primarily achieved during glacial maxima, but during periods of
much more limited glacierization, when these areas merely maintained local ice-caps and valley glaciers and when high local ablation
gradients produced rapid local glacier movement.

One of the major centres of ice outflow, that of the Southern
Uplands of Scotland, shows a lesser erosional intensity than the
surrounding areas. We would suggest that this arises because of its
lower elevation compared with the other centres of outflow, which
did not enable it to support as much active local ice during the

periods when no large British ice-sheet existed. It suffered less erosion during these periods, and the smaller erosional intensity that it shows compared with surrounding areas is a product of this and low erosional rates beneath source areas during periods of more extensive ice-sheet glaciation. The absence of erosional features in Buchan (north-east Scotland), a lowland area, similarly reflects lack of erosion beneath the central part of the ice-sheet, and the absence of hills to support more active local ice during periods of less extensive ice cover. Although we have no general picture of the intensity of glacial erosion in Ireland, we would predict that the low-lying source area in central Ireland will show a much lesser intensity than the marginal areas for reasons similar to those which we suggest for Buchan. We also would suggest that the deeply weathered soils found in Aberdeenshire, beneath the Scottish ice-sheet source, also reflect lack of erosion beneath a sluggish, possibly cold glacier sole. The deeply scoured valleys of the region again reflect erosion by valley glaciers and local ice-caps.

In the previous section we drew attention to the intensified erosion near to the margins of ice-sheets in lowland England. We would suggest that this reflects the higher velocities and temperate basal conditions which the model predicts will occur there, both of which are likely to enhance erosion.

Patterns of glacial deposition

Fig. 17.4 is a schematic diagram of the distribution of glacial deposits over England, Wales, and Scotland. Within the area covered by the Late Devensian ice-sheet we have distinguished areas in which drumlins occur (including Ireland) and areas which we consider to be areas of supraglacial deposition.

The majority of British drumlins are composed of till. Where broad areas of subglacially deposited lodgement till are exposed beyond modern glaciers it is almost invariably found that the surface is moulded into drumlins if the till is thick, or into more muted drumlinoid forms if the till is thin. Thus we suggest that the areas of drumlins in Fig. 17.4 represent areas of thick subglacial till (or a subglacial surface) which has not subsequently been covered by supraglacial deposits or proglacial outwash. Lodgement tills occur in other areas, but we would suggest that they are generally too thin to have developed drumlins large enough to survive subsequent disturbing influences, or that drumlins have been covered by other sediments.

The areas which we characterize as areas of supraglacial deposition show a hummocky morainic and kamiform topography which we believe to reflect deposition on and between stagnant ice masses.

Such topographic forms develop most extensively at the margins of modern glaciers where thick englacial debris sequences occur. During glacier retreat this debris becomes exposed as till on the glacier surface in the terminal area, it inhibits ablation of the underlying ice, and leads to the formation of extensive areas of till-covered stagnant ice which grows by proximal accretion as the active glacier margin retreats. Glaciofluvial and glaciolacustrine outwash sediments accumulate on and between these till-covered stagnant ice-masses and when buried ice melts, leave a hummocky topography. This hummocky topography may be predominantly composed of melt-out till and flow till; or of lacustrine and fluvial sediments showing gently folded bedding structures which reflect the surface topography, and faults; or a combination of till and outwash sediments (Boulton 1972*a*). Boulton and Paul (1976) have suggested the terms 'supraglacial land system' and 'supraglacial sediment association' to describe such areas. The thick englacial debris sequences in ice-sheets, which favour the development of the supraglacial land system and sediment association, are generally thought to be incorporated within the glacier where there is net freezing of water to the glacier sole (Weertman 1961*b*; Boulton 1972*b*).

A similar association is also produced in valley glaciers where medial and lateral moraines form englacial debris sequences (glaciated valley land system, Boulton and Paul 1976) but this does not carry the same implications as in lowland areas of ice-sheet glaciation. Much of the supraglacial sediment shown in Fig. 17.4 in highland Britain has this origin.

Thus, we believe that Fig. 17.4 distinguishes broadly between areas where subglacial lodgement till deposition followed by proglacial outwash deposition have been dominant (subglacial/proglacial land system and sediment association) and areas where ice stagnation and supraglacial deposition have been dominant. In the former areas debris was transported in a thin basal horizon as in modern temperate glaciers, and deposited subglacially as till; in the latter areas it was transported as thick englacial debris sequences, as in modern polar and subpolar glaciers, and much was deposited as flow till and melt-out till.

An explanation of observed depositional patterns

Vernon (1966) has shown that the main drumlin fields in Ireland are located at some distance behind the margins of the Late Devensian ice-sheet. The same appears to be true in England, although they are found within 10 km of the eastern margin of the Welsh ice-cap.

We would therefore conclude that conditions for the formation of recognizable drumlins did not occur beneath the marginal zone of the Late Devensian ice-sheet at its maximum. Smalley and Unwin (1968) have proposed a mechanism for drumlin formation which depends on the dilatant behaviour of till, in which they suggest that drumlins will only form over a limited range of hydrostatic stress, and that the lower stress levels near to the margins of ice-sheets will inhibit this mechanism. According to our model, drumlins in England do not occur in the marginal zones of the ice-sheet where the ice is less than about 800 m thick, equivalent to an ice pressure of about 80 bar, more than enough to inhibit any deformation in subglacial sediments when the average shear stress at the bed of a glacier is rarely more than 1 bar. Of course, this ignores a subglacial water pressure which may very much reduce the effective hydrostatic stress, but as has been shown (Boulton 1975), the value of this hydrostatic stress is not directly related to ice thickness; thus we would suggest that Smalley and Unwin's (1968) hypothesis cannot explain the absence of drumlins in the marginal areas of the Late Devensian ice-sheet, especially in view of the fact that drumlins can form under very small ice thicknesses in present-day glaciers.

We wish to suggest an alternative hypothesis. Without wishing to speculate on the mechanisms of drumlin formation, we suggest as a matter of observation that drumlins commonly develop on accumulating lodgement till surfaces at the margins of modern glaciers. We also suggest that the presence or absence of lodgement till beneath and beyond the margins of modern glaciers depends on the glacier sliding velocity. Where sliding velocities do not exceed the order of 50–100 m/year it is found that there is a partial or complete lodgement till cover on the subglacial bed, whereas at higher sliding velocities the subglacial bed is actively eroded and swept clear of any lodgement till.

It was suggested in a previous section that the greater intensity of erosion postulated for the marginal area of the ice-sheet in England was a result of high ice velocities. The lower velocities on the up-glacier side of this marginal zone may have allowed lodgement till deposition to occur for a sufficiently long time for large drumlins to form on its surface. The absence of drumlins down the east coast of England may also reflect the high ice velocities inferred for that area, whilst the presence of drumlins near to the Welsh ice-cap margin probably reflects the low velocities likely in that ice-cap. At the extreme margin of a steady-state glacier, the ice velocity is equal merely to the ablation rate. Thus in our model, the velocities shown in Fig. 17.2 should decrease rapidly to very low values at the extreme margin, and therefore beyond a marginal high-velocity zone of erosion we expect an extreme marginal zone of deposition.

If we consider an ice-sheet advancing to its maximum extent and subsequently retreating, we would expect a broad marginal zone of erosion to sweep forward to the maximum extent of the ice-sheet. As this stabilized, we would expect deposition of lodgement till at the extreme margin. During the early stages of retreat, the extreme marginal zone would deposit a thin stratum of lodgement till as it moved north, and although it may have had drumlinoid forms on its surface, the limited period of deposition would ensure that these were not sufficiently strongly sculptured to survive the ensuing millenia as recognizable forms. In the internal zones of lower velocity however, there was time enough for thick till and well-defined drumlins to form. We assume that the climatic amelioration which caused retreat of the ice sheet would have reduced velocities and produced general lodgement till deposition and drumlin formation. Thus most of the drumlins that we find further north may well have had their final form imposed near to the retreating ice-sheet margin (see Embleton and King 1975). Such a hypothesis might also explain Godlthwait's (1974) suggestion that near the southern margin of the Laurentide ice-sheet, a final phase of deposition succeeded an earlier phase of erosion.

It was argued above that the supraglacial sediment association is most likely to have been derived from a glacier with a thick sequence of englacial debris, and that in an ice-sheet such sequences are probably incorporated as a result of net freezing of subglacial water on to the glacier sole. We have suggested that such a situation might occur where the British ice-sheet, with an internal zone of freezing and an outer zone of melting, moved over permafrost so that the extreme margin of the ice-sheet was cooled, thus giving rise to the supraglacial sediment association, which in an area such as the Cheshire–Shropshire basin far exceeds in volume the underlying subglacially deposited materials. A similar sequence, passing from a marginal suite of supraglacially-deposited sediments lying above a subglacial horizon of predominant erosion and little deposition, to an inner suite reflecting predominant subglacial deposition, has been described by Clayton and Moran (1974) from the Great Plains of North America. It should be stressed, however, that the sedimentational and erosional model developed here applies only to the period of the Late Devensian glacial maximum, and that the suites of sediments and landforms in more central areas must be related to the very different dynamic and thermal structures which probably held during the retreat phases of the British ice-sheet.

Acknowledgement

F. Grynkewicz assisted with the computations.

References

Boulton, G.S. (1972*a*). The role of thermal regime in glacial sedimentation. *Polar Geomorphology, Spec. publ. Inst. Geogr.* 4.

Boulton, G.S. (1972*b*). Modern Arctic glaciers as depositional models for former ice sheets. *J. Geol. Soc. Lond.* **128**, 361-93.

Boulton, G.S. (1974). Processes and patterns of glacial erosion. In *Glacial Geomorphology* (ed. D.R. Coates). New York State University, Binghamton.

Boulton, G.S. (1975). Processes and patterns of subglacial sedimentation: a theoretical approach. In *Ice Ages: Ancient and Modern.* (ed. H.E. Wright and F. Mosley). *Geol. J. Spec. Issue.* Liverpool.

Boulton, G.S. and Paul, M.A. (1976). The influences of genetic processes on some geotechnical properties of subglacial tills. *Q. J. Eng. Geol.* **9**, 154-194.

Bromley, R.G. (1967). Field meeting on the chalk of Cambridgeshire and Hertfordshire. *Proc. Geol. Ass.* **77**, 277-9.

Charlesworth, J.K. (1957). *The Quaternary Era.* Edward Arnold, London.

Clayton, K.M. (1974). Zones of glacial erosion. *Spec. Publ. Inst. Geogr.* **7**, 163-76.

Clayton, L. and Moran, S.R. (1974). A glacial process form model. In *Glacial Geomorphology* (ed. D.R. Coates). New York State University, Binghamton.

Coope, G.R. (1975). Climatic fluctuations in north-west Europe since the last Interglacial indicated by fossil assemblages of Coleoptera. In *Ice Ages: Ancient and Modern* (ed. H.E. Wright and F. Mosley). *Geol. J. Spec. Issue*, **6**, Liverpool.

Embleton, C. and King, C.A.M. (1975). *Glacial geomorphology.* Edward Arnold, London.

Farrington, A. (1953). Local Pleistocene glaciation and the level of the snow line of Croaghaun Mountain, in Achill Island, Co. Mayo, Ireland. *J. Glaciol.* **2**, 262-7.

Goldthwait, R.P. (1974). Till deposition versus glacial erosion. In *Research in Polar and Alpine geomorphology, 3rd Guelph Symposium in Geomorphology* (eds. B.H. Fahey and R.D. Thompson) Geo. Abstracts, Norwich. 159-166.

Hoppe, Gunnar, (1974). The glacial history of the Shetland Islands. In *Progress in geomorphology, papers in honour of David L. Linton, Spec. Publ. Inst. Geogr.* **7**, 197-210.

John, B.S. (1970). Pembrokeshire. In *The glaciations of Wales and adjoining regions* (ed. C.A. Lewis). Longman, London.

Lamb, H.H. and Woodroffe, A. (1970). Atmospheric circulation during the last ice age. *Quaternary Res.* **1**, 29-58.

Manley, G. (1959). The Late-Glacial climate of north-west England. *Lpool and Manchr. geol. J.* **2**, 188-215.

Smalley, I.J. and Unwin, D.J. (1968). The formation and shape of drumlins and their orientation in drumlin fields. *J. Glaciol.* **7**, 377-90.

Synge, F.M. (1969). The Würm ice limit in the West of Ireland. In *Quaternary Geology and Climate* (ed. H.E. Wright). Proc. VII Congress INQUA, Vol. 16.

Tricart, J. (1970). *Geomorphology of cold environments.* Macmillan, London.

Vernon, P. (1966). Drumlins and Pleistocene ice flow over the Ards peninsula–Strangford Lough area, County Down, Ireland. *J. Glaciol.* **6**, 401-9.

Weertman, J. (1961*a*). Stability of ice-age ice sheets. *J. geophys. Res.* **66**, 3788–92.

Weertman, J. (1961*b*). Mechanism for the formation of inner moraines found near the edge of cold ice caps and ice sheets. *J. Glaciol.* **3**, 965–78.

Weertman, J. (1972). General theory of water flow at the base of a glacier or ice sheet. *revs. Geophys. Space Phys.* **10**, 287–333.

Williams, J. and Barry, R.G. (1974). Ice Age experiments with the NCAR General Circulation Model: conditions in the vicinity of the northern continental ice sheets. In *Climate of the Arctic.* (ed. G. Weller and S.A. Bowling). Geophysical Institute, Fairbanks, Alaska.

Williams, R.B.G. (1975). The British climate during the Last Glaciation; an interpretation based on periglacial phenomena. In *Ice Ages: Ancient and Modern.* (ed. H.E. Wright and F. Mosley). *geol. J. Spec. Issue* **6**, Liverpool.

18

Preglacial Pleistocene deposits of East Anglia

B. M. FUNNELL and R. G. WEST

Abstract

The preglacial Pleistocene deposits of East Anglia are allocated to three formations: the Red Crag Series, the Norwich Crag Series, and the Cromer Forest Bed Series. Constituent members and beds are listed.

Beds, members, and foraminiferal and molluscan assemblages and assemblage zones are provisionally allocated to a pollen stage system consisting of the following stages: pre-Ludhamian, Ludhamian, Thurnian, Antian, Baventian, Pastonian, Beestonian, and Cromerian. Problems of correlation to this sequence of stages are stated.

Introduction

Many of the difficulties and misunderstandings in the interpretation of the stratigraphy of the early Pleistocene deposits of East Anglia result from failure to apply modern stratigraphical procedures. For this reason we here review these deposits within a standardized stratigraphical framework, giving just sufficient information on earlier stratigraphies to enable them to be related to our own modern shce scheme. In this way we hope to present a coherent view of the present state of knowledge, and to establish a clear context for future advances. We first construct a comprehensive lithostratigraphic classification of beds, members, and formations, and then consider their bio-stratigraphic or chronostratigraphic attrubution in terms of zones and stages.

The early Pleistocene deposits of East Anglia have long since been allocated to two major Series, corresponding to formations in modern usuage (George *et al.* 1969). These are the Norwich Crag Series (Woodward 1881) and the Cromer Forest Bed Series (Reid 1882). We propose to add a third, the Red Crag Series. The constituent members of these formations are shown in Table 18 1. Locations are given in Fig. 18.1.

Preglacial formations, their lithology and distribution

The principal preglacial formations (Table 18.1) are described in approximate stratigraphical order.

Red Crag Series

This formational name is applied to shelly marine sands, containing minor amounts of clay and gravel, occurring at the surface and at depth in east Norfolk, east Suffolk, and north-east Essex. Grey in colour at depth, they are deep red at outcrop due to oxidation. Large-scale cross-bedding, indicating deposition as submarine dunes (or sand waves) is characteristic of surface exposures, although often the formation is terminated upwards by a phase of horizontal bedding.

The Red Crag member. This was divided by Harmer (1900) into a series of local subdivisions: 'Waltonian', 'Newbournian', 'Butleyan', on geographical and faunal grounds. These subdivisions are not laterally continuous and cannot be sustained on lithological grounds. We suggest discontinuation of these stage-like names in favour of geographical lithostratigraphic expressions Walton, Newbourn, and Butley Crags. The Red Crag occurs in east Suffolk and north-east Essex only. It is typically 10 m thick, but reaches a maximum of about 25 m at Stradbroke.

Table 18.1. *Components of the preglacial formations of East Anglia*

Members	Beds
A. *Cromer Forest Bed Series/Formation*	
Bacton Member (freshwater)	Arctic Freshwater Bed (part of Reid's 'Upper Freshwater Bed')
Mundesley Member (marine)	*Yoldia* [*Leda*] *myalis* Bed Mundesley Clay (part of Reid's 'Forest Bed (estuarine)')
West Runton Member (freshwater)	West Runton Freshwater Bed (Reid's 'Upper Freshwater Bed' at West Runton) Corton Rootlet Bed
Runton Member	Woman Hithe Gravel Woman Hithe Clay
Paston Member (marine)	West Runton Clay (part of Reid's 'Forest Bed (estuarine)')
Sheringham Member (freshwater)	(Reid's 'Forest Bed' at Happisburgh, and 'Upper Freshwater Bed' at Beeston)

'Weybourne Crag' facies

B. *Norwich Crag Series/Formation*	
Norwich Crag Member	Westleton Beds (*sensu* Hey 1967) Easton Bavents Clay
Chillesford Beds Member	Chillesford Clay Chillesford Crag *Scrobicularia* Crag

C. *Red Crag Series/Formation*	
Ludham Crag Member	
Red Crag Member	Includes: Butley Crag (= Butley horizon of Harmer) Newbourn Crag (= Newbourn horizon of Harmer) Walton Crag (= Walton and Little Oakley horizons of Harmer)

Ludham Crag (Funnell 1961*b*). This name was given to 18 m of coarse shelly sands at the base of a borehole in east Norfolk. For further details of its lithology see Funnell (1961*b*) and West (1961).

Norwich Crag Series (Woodward 1881)
This formation contains three members, the Norwich Crag *sensu stricto*, a marine sand with shell beds, clay beds, and gravelly horizons, occurring in the vicinity of Norwich, throughout east Norfolk and into north-east Suffolk; the Chillesford Beds of east Suffolk, and

the Westleton Beds, a distinctive gravel horizon in south-east Norfolk and north-east Suffolk (Hey 1967).

Norwich Crag (Woodward 1881). This member reaches a thickness of about 18 m at Ludham and 12 m at the type locality of Bramerton (Funnell 1961*b*). It consists of yellow-brown sands (grey at depth), sometimes micaceous, occasionally with shells, and layers of clay. Where the member rests directly on the Chalk there is usually a basement bed of large flints. Elsewhere rounded flint gravels are developed.

Chillesford Beds (Prestwich 1871). Sub-divisible into three: the *Scrobicularia* Crag (Wood 1886) below, yellow-brown sands and shelly

Fig. 18.1. Preglacial Pleistocene localities of East Anglia

sands containing a significant content of derived Red and Coralline Crag fossils and limited to the Chillesford–Sizewell area; the Chillesford Crag, a shelly marine sand; and the Chillesford Clay, a marine intertidal silty clay above. The geographical extent of the upper two beds was greatly exaggerated by nineteenth century investigators. The true occurrence is restricted to the immediate vicinity of Chillesford (Funnell 1961*b*; West and Norton 1974).

Cromer Forest Bed Series (Reid 1882).
This formation consists of both freshwater and marine sediments. It is exposed around the Norfolk coast from Weybourne to Happisburgh and on the Suffolk coast from Hopton to Kessingland, reaching a maximum thickness of about 8 m. West of Cromer the formation lies between the Chalk and the Anglian glacial deposits. To the southeast it lies on marine Norwich Crag.

The following well-established members and beds do not fully cover the great variety of sedimentary units which comprise the Cromer Forest Bed Series.

Weybourne Crag (Reid 1882). This is probably not a true member, but a shelly marine facies developed at more than one horizon, usually near the base of the Series and on the Chalk.

Sheringham Member. Impersistent horizons of organic mud up to about 1 m thick occur at several Norfolk coastal localities. The member includes the 'Forest Bed' at Happisburgh and Reid's 'Upper Freshwater Bed' at Beeston.

Paston Member. This consists of up to about 3 m of tidal silts and clays and marine shelly sands. The *West Runton Clay*, 1·5 m of tidal silty clay, may be distinguished as a component bed.

Runton Member. Up to about 3 m thick, this is composed of beach gravels, sometimes shelly above and freshwater silty clays below. These two component beds of the Runton Member are designated the *Woman Hithe Gravel* and the *Woman Hithe Clay*.

West Runton Member. This member includes layers of alluvial clay and organic mud up to about 1·5 m thick, exposed on the Norfolk and Suffolk coasts. Two beds may be distinguished: the *West Runton Freshwater Bed* ('Upper Freshwater Bed' of Reid) and the *Corton Rootlet Bed*.

Mundesley Member. This includes tidal silty clays and beach gravels and sands, occurring on the Norfolk and Suffolk coasts and reaching a thickness of up to about 2·5 m. The *Yoldia* [*Leda*] *Myalis Bed* of Reid (1882) at West Runton, and the *Mundesley Clay*, a prominent bed of tidal, laminated silty clay at the base of the Mundesley–Paston cliffs, may be distinguished as component beds of this member.

Bacton Member. This member includes clays and organic muds, up to about 1·5 m thick, exposed on the Norfolk coast. It includes the *Arctic Freshwater Bed* of Reid (1882), and other organic horizons included in Reid's 'Upper Freshwater Bed'.

Pollen Stages

A division of the preglacial Pleistocene can be made on the basis of the sequence of pollen assemblage biozones. The principal evidence comes from the Ludham borehole (West 1961), the Cromer Forest Bed Series (West unpublished) and the Crag of south-east Suffolk (West and Norton 1974). The biozones distinguished in these sequences are now described.

Ludham borehole

Lp 5 (Pastonian?)	Biozone of *Betula, Pinus, Picea, Quercus, Alnus,* and *Carpinus.* AP > NAP†.
Lp 4 (Baventian)	Biozone of *Betula, Pinus, Picea* and *Alnus.* NAP>AP. Gramineae and Ericales with high values.
Lp 4c	*Pinus* dominant in AP.
Lp 4b	*Betula, Pinus,* and *Alnus* equally frequent.
Lp 3 (Antian)	Biozone of *Betula, Pinus, Picea, Tsuga, Ulmus, Quercus, Alnus, Carpinus,* and *Pterocarya.* AP>NAP.
Lp 2 (Thurnian)	Biozone of *Pinus* (dominant), *Betula, Picea,* and *Alnus;* low frequencies of *Tsuga.* NAP>AP. Gramineae and Ericales with high frequencies.
Lp 1 (Ludhamian)	Biozone of *Pinus* (dominant), *Betula, Picea, Tsuga, Ulmus, Quercus, Alnus, Carpinus,* and *Pterocarya.* AP> NAP.
Lp 1b	High frequencies of *Tsuga.*
Lp 1a	Low frequencies of *Tsuga.*

Cromer Forest Bed Series

Pollen assemblage biozones from this formation and its associated underlying and overlying sediments have been distinguished and grouped in stages as follows:

---

†AP, arboreal pollen; NAP, non-arboreal pollen.

Stage		Biozone
Anglian	e An	Biozone of high NAP, Gramineae.
	Cr IV	Biozone of *Pinus, Picea, Betula*, and *Alnus*.
	Cr III	Biozone of *Carpinus, Abies*.
Cromerian	Cr II	Biozone of *Pinus, Quercus, Ulmus*, and *Tilia*.
	Cr I	Biozone of *Pinus, Betula*.
Beestonian	Be	Biozone of high NAP, Gramineae.
	Pa IV	Biozone of *Pinus, Picea, Betula*, and *Ericales*.
Pastonian	Pa III	Biozone of *Pinus, Betula, Picea*, and *Carpinus*.
	Pa II	Biozone of *Pinus, Quercus, Ulmus*, and *Carpinus*.
	Pa I	Biozone of *Pinus, Betula*.
	Pre Pa d	Biozone of Gramineae, Cyperaceae, *Betula*, and *Pinus*.
	Pre Pa c	Biozone of Gramineae, *Pinus, Betula*, and *Alnus*.
Pre-Pastonian	Pre Pa b	Biozone of Gramineae, *Artemisia*.
	Pre Pa a	Biozone of *Pinus, Betula*, Gramineae, and Ericales.

East Suffolk

West and Norton (1974) identified a Chillesford pollen assemblage of temperate type (biozone of *Pinus, Quercus, Alnus, Carpinus*, and *Picea*) from the Chillesford Crag and correlated crag at Aldeburgh, Sizewell, Thorpe Aldringham, and Wangford.

Other Sites

A pollen assemblage biozone characterized by very high frequencies of *Pinus* and low NAP was found below a biozone of Ludhamian type in the Stradbroke, Suffolk, borehole (Beck, Funnell, and Lord 1972). A similar type of assemblage was found at depth in samples of crag from Sizewell (West and Norton 1974). This biozone has been included in the pre-Ludhamian stage.

At Easton Bavents (Funnell and West 1962) two pollen assemblage biozones are present. The upper is very similar to Lp 4 at Ludham, with high NAP (Gramineae and Ericales), and they were therefore both included in the Baventian stage. The lower, with *Tsuga*, was correlated with the Antian.

Synthesis of stages based on pollen assemblage biozones

The biozones described above can be grouped as a succession of stages as shown in the following table.

Ludham borehole	Cromer Forest Bed Series	East Suffolk	Stradbroke borehole
	Anglian (c, g)		
	Cromerian (t)		
	Beestonian (c)		
Lp 5 Pastonian (t) -?-	Pastonian (t) -?-	Chillesford p.a.b. (t)	
Lp 4 Baventian (c) -?-	Pre-Pastonian (c)-?-	Baventian (c) of Easton Bavents	
Lp 3 Antian (t)		Antian (t)	
Lp 2 Thurnian (c)			
Lp 1 Ludhamian (t)			Ludhamian (t)
			Pre-Ludhamian

(g, glacial; c, cold; t, temperate; p.a.b., pollen assemblage biozone).

The principal problems in this synthesis are as follows:

(i) Is the pre-Pastonian of the Norfolk coast to be correlated with the Baventian of the Ludham borehole? The answer is probably yes. The pre-Pastonian is well developed at Happisburgh, 7 miles north of Ludham, and is at the same depth O.D. as the Baventian at Ludham.

(ii) Is the Chillesford p.a.b. to be correlated with the Pastonian? This correlation was tentatively made by West and Norton (1974), but it may be that this assemblage belongs to an older temperate stage, since if the correlation is correct we have to explain the absence of *Macoma balthica* from the Suffolk crag, compared with its abundance in the pre-Pastonian and later stages of the Norfolk coast. If the correlation is incorrect, the Suffolk crag with the Chillesford pollen assemblage will belong to an older temperate stage, perhaps part of the Antian or a hitherto unrecorded separate temperate stage later than the Thurnian.

(iii) Is the Baventian type of assemblage with high NAP (Gramineae and Ericales) a recurrent facies assemblage? Is the Baventian at the type site securely correlated to Lp 4 at Ludham?

These questions can only be answered by further study.

Faunal stages and their stratigraphical interpretation

The marine (and freshwater) molluscs of the East Anglian preglacial deposits.

These have been studied for a long time. Classical studies were undertaken in the late nineteenth and early twentieth centuries by Wood (1848–1882) and Harmer (1900, 1914–1925). Modern quantitative studies have been undertaken by Norton (1967, 1970; in Norton and Beck 1972; in West and Norton 1974).

The *Foraminifera* were also studied and monographed in the nineteenth century; recent quantitative work has been undertaken by Funnell (1961*b*; in Funnell and West 1962; in Beck *et al.* 1972).

Neither of these two groups has yet been used to provide a comprehensive bio-, or time-stratigraphic classification of the East Anglian preglacial, although parts of a classification exist. A major reason for this is the strong facies control to which they are subject, which makes them respond more to changes in local environment than to longer-term changes. In these circumstances, much more reliance is placed on the often less-abundant pollen content of these sediments to erect biozones and stages (q.v.)

Harmer's (1900) mollusc 'zones'

Harmer (1900) postulated eight mollusc 'zones' as follows:

Zone of *Tellina balthica* = Weybourne Crag facies of the Cromer Forest Bed Series.

Zone of *Leda oblongoides* = Chillesford Crag Bed of the Chillesford Beds Member in the Norwich Crag Series.

Zone of *Astarte borealis* = Norwich Crag Member of the Norwich Crag Series in Norfolk (called the *Upper* Division of the Norwich Crag by Harmer).

Zone of *Mactra subtruncata* = Norwich Crag Member of the Norwich Crag Series in north Suffolk (called the *Lower* Division of the Norwich Crag by Harmer).

Zone of *Cardium groenlandicum* = Butley Crag of the Red Crag Series (Harmer's 'Butleyan').

Zone of *Mactra constricta* = Newbourn Crag of the Red Crag Series (Harmer's 'Newbournian').

Zone of *Mactra obtruncata* = Oakley horizon of the Walton Crag of the Red Crag Series (part of Harmer's 'Waltonian').

Zone of *Neptunea contraria* = Walton horizon of the Walton Crag of the Red Crag Series (part of Harmer's 'Waltonian').

Harmer's mollusc 'zones' cannot be regarded as biozones in the modern sense. They are in effect little more than palaeontological labels of convenience for beds, members, horizons, and facies. The same can be said of Harmer's 'stages' (Waltonian, Newbournian, Butleyan, Icenian, Chillesfordian, Weybournian), which often have a strong geographical rather than stratigraphical connotation. It is recommended therefore that the use of Harmer's zones and stages be discontinued and a more informal nomenclature adopted, linked to geographical locations, as in this chapter.

The removal of these molluscan pseudo-zones and pseudo-stages from the literature should lead to freer discussion of the molluscan (and foraminiferal) successions at the North Sea margins. Contemporaneous pollen biozones/stages are now available for practically the entire period under consideration, with the possible exception of the pre-Ludhamian (for which Waltonian was temporarily retained by Mitchell *et al.* (1973)).

Stratigraphical distribution of the Mollusca

Norton (1967) recognized 6 molluscan assemblage biozones in the Ludham succession. These relate to the pollen stages as follows:

Lm 6†	=	Lp 4b	Baventian
		Lp 3	Antian
Lm 5	=	Lp 3	Antian
		Lp 2	Thurnian
Lm 4	=	Lp 2	Thurnian
Lm 3	=	Lp 1b	Ludhamian
Lm 2		Lp 1b	Ludhamian
		Lp 1a	
Lm 1b		Lp 1a	Ludhamian
Lm 1a			

Norton (1967) also recognized molluscan assemblage zones at Bramerton: Bm 1 and Bm 2. Further details of the molluscs contained in these biozones, and their relation to the pollen stages, are given in the next section of this chapter and in Chapter 4.

Since 1967, Norton (1970; Norton and Beck 1972; West and Norton 1974) has not proposed any further molluscan assemblage biozones, but has interpreted molluscan assemblages from Aldeburgh, Aldeby, Chillesford, Easton Bavents, Sizewell, Thorpe Aldringham, and Wangford in terms of marine facies, without attributing any zonal value to these assemblages. The apparent stratigraphical position of these assemblages and their relation to molluscan biozones proposed for the Netherlands is considered in the next section of this chapter.

Stratigraphical distribution of the Foraminifera

Funnell (1961*b*) proposed a series of 7 horizons, based on foraminifers, for the Ludham succession and a further 3 horizons (Bf 1, Bf 2, Bf 3) for the Bramerton succession. There seems to be no significant difference between these horizons and the assemblage biozones of Norton (1967) and they are therefore now called assemblage biozones. At Ludham they relate to the pollen stages as follows:

†Originally rendered as L.M.6, etc. In this chapter faunal and floral biozones are consistently presented conforming to the pattern: initial capital letter(s) indicating locality; lower case letter indicating biological group, and *arabic* numeral indicating number of biozone.

Lf 7†	=	Lp 4c	Baventian
Lf 6	=	Lp 4b	Baventian
Lf 5	=	Lp 3	Antian
		⎰?Lp 2	?Thurnian
Lf 4	=	⎱ Lp 2	Thurnian
Lf 3	=	Lp 1b	Ludhamian
Lf 2	=	Lp 1a	Ludhamian
Lf 1	=	Lp 1a	Ludhamian

Since 1961 Funnell (Funnell and West 1962) has described foraminiferal assemblage biozones (≡ 'horizons') from Easton Bavents (EB f 1 and EB f 2). He has also described assemblages, without referring them to assemblage biozones, from Chillesford, Sidestrand, Sizewell, Thorpe Aldringham (Funnell 1961*b*) and from Stradbroke, Walton, Ipswich, Newbourn, and Butley (in Beck *et al.* 1972, and Table 18. 2 and Fig. 18. 2). The apparent stratigraphical position of these assemblages, and their relation to foraminiferal biozones proposed for the Netherlands are considered next.

RED CRAG SERIES FORAMINIFERA

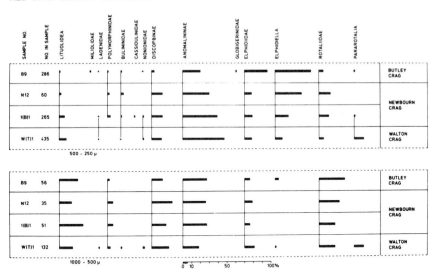

Fig. 18. 2. Red Crag Series Foraminifera

Stratigraphical and general environmental interpretation

The sequence of preglacial deposits under consideration can currently be ranged under eight stages, based on pollen spectra (see above). Each of these stages is now considered in turn.

---

†originally rendered as LVII, etc. See footnote on p. 256.

Table 18. 2. *Red Crag Series Foraminifera*

Size Range	500-250μm				1000-500μm			
Sample No.	W(T)1	1(B)1	N12	B9	W(T)1	1(B)1	N12	B9
No. in Sample	435	265	60	286	132	51	35	56
Buccella inusitata	9	7	5	0	3			
Bulimina aculeata		0	3	1				
Bulimina	0	0		0	1			
Cassidulina laevigata var. *carinata*		1						
Cibicides refulgens	1							
C. lobatula	18	20	10	16	2			12
C. lobatula var. *grossa*	6	4	13	2	6	4	3	10
C. pseudoungeriana	14	6	3	1	1			
C. subhaidingerii	7	9	3	1	9	24	23	6
Dorothia gibbosa var. *alleni*		0				2		2
Elphidiella hannai	8	15	30	41	1			4
Elphidium clavatum	2	+	2	7				
E. crispum	1	2			11	6	9	6
E. frigidum		0		9				
E. haagensis	3	1	5	0				
E. macellum var. *granulosum*	0	4		1				
E. orbiculare				0				
E. pseudolessonii	0	0	2	8				
E. selseyensis		0						
Elphidium	1	1						
Eponides repandus	0	3	5	2	16	16	23	12
Faujasina subrotunda	0	0						
Gaudryina (Siphogaudryina) tumidula	1	2	2		15	22	14	17
Globigerina bulloides				1				
Globulina inaequalis		0						
G. aff. myristiformis		2	2	1				
Globulina						2		
Guttulina lactea		0						
Lenticulina	0			0	1			
Nodosaria (?)	0							
Nonion lamarcki	0							
Nonion	0	0		0	1			
Nonionella janiformis	0	0		0	1			
Pararotalia serrata	11	1		1	11			
Planularia		0						
Polymorphina					1		6	
Pseudopolymorphina		1			2	4		2
Pullenia sphaeroides	0	0						
Quinqueloculina				1				

Size Range	500–250µm				1000–500µm			
Sample No.	W(T)1	1(B)1	N12	B9	W(T)1	1(B)1	N12	B9
No. in Sample	435	265	60	286	132	51	35	56
Rosalina parisiensis	2	2	2	0				
Rosalina				0				
'Rotalia' beccarii	5	11	13	5	18	18	23	29
Textularia sagittula	5	3	1					
T. suttonensis	0	0						
T. truncata	1	1						2
Textularia	0							
Textulariella trochoides		0				4		

Pre-Ludhamian Stage

The first stage recognized in the Pleistocene deposits of East Anglia is the pre-Ludhamian (Beck, Funnell, and Lord 1972). The pollen of this stage is dominated by *Pinus*, associated with up to 5% *Betula*, *Picea*, and *Alnus* in total; Ericales 5–10% and Gramineae 5%. It occurs between −15 and −40 m OD at Stradbroke in Suffolk (type locality), and has also been recognized at between −36 and −37 m OD at Sizewell, Suffolk (West and Norton 1974). The pollen indicates the regional presence of coniferous woodland of a cool temperate type.

In the lower part of the stage at the type site the foraminifers (Beck *et al.* 1972) closely resemble those of the Walton Crag (Fig. 18.2 and Table 18.1, from Funnell 1961*a*, Diagram 11 and Table III). *Pararotalia serrata* is abundant, and *Elphidium haagensis* is present, unaccompanied by *Elphidium frigidum*. In the upper part of the stage at the type site, the foraminifers resemble those of the Butley Crag. *P. serrata* and *E. haagensis* are much reduced or absent, whereas *E. frigidum* is consistently present. There is also a significant influx of the planktonic foraminifers *Globigerina bulloides* and *Globigerina pachyderma*, both in the upper part of the pre-Ludhamian stage at Stradbroke and in the Butley Crag. The assemblages with *P. serrata* suggest Lusitanian affinities, and warm temperate conditions, explicable if the North Sea was broadly connected to the Atlantic via the English Channel region. On the other hand, the assemblages with *E. frigidum* indicate Boreal affinities with cool temperate conditions, and the planktonic foraminifers may imply the injection of oceanic water into the North Sea principally around the north of Scotland. The foraminifers of the Sizewell deposits have not yet been examined.

Mollusca from the pre-Ludhamian at Stradbroke and Sizewell have not yet been examined. Lists were given by Harmer (1900) for the Walton, Butley (and intervening Little Oakley and Newbourn) Crags; see also West and Norton (1974; p. 19, table 1, sample Chillesford 8).

Mollusc assemblages from the Walton Crag closely resemble those of the Sande von Merksem of Belgium (see Zagwijn 1974), which are attributed to the *Nassarius propinquus–Lentidium complanatum* assemblage zone (=MOL.C) of the modern Netherlands mollusc zonation (Spaink 1975).

Mollusc assemblages from the Butley Crag resemble those of the *Serripes groenlandicus–Yoldia lanceolata* assemblage zone (=MOL.B) of the Netherlands zonation (Zagwijn 1974; Spaink 1975).

The sediments of the pre-Ludhamian stage at Stradbroke are normally magnetized (van Montfrans 1971; Beck *et al.* 1972).

Ludhamian Stage

The second stage recognized in the Pleistocene deposits of East Anglia is the Ludhamian (West 1961). The pollen spectra comprise *Pinus, Betula, Picea, Tsuga, Ulmus, Quercus, Alnus, Pterocarya*, and Cupressaceae; arboreal pollen is much more abundant than non-arboreal pollen and Chenopodiaceae are frequent. In addition to the type site, this stage has also been recognized at Stradbroke, Suffolk, between +30 and −15 m OD (Beck *et al.* 1972). The pollen suggests a temperate forest cover characterized by the widespread presence of *Tsuga* and *Pterocarya*.

The foraminifers of the lower part of the Ludhamian (Lp 1a) at the type locality (Funnell 1961*b*, samples 20–32) are similar to, but more restricted than, those in the lower part of the pre-Ludhamian and the Walton and Newbourn Crags. The resurgence of *P. serrata*, now consistently accompanied by *E. frigidum*, is notable. The upper part of the Ludhamian (Lp 1b) at the type locality contains a distinctly different and clearly reduced foraminiferal fauna, similar to those in the succeeding Norwich Crag. It is for this reason, and the accompanying lithological changes, that Funnell (1961*b*) proposed the name Ludham Crag for the beds of the lower part of the Ludhamian. The foraminifers of the upper Ludhamian at Ludham are dominated by *Elphidiella hannai* and other Elphidiidae; *Elphidium excavatum* and *Rotalia perlucida* are indicative of a shallow water, estuarine or bay-head environment, whilst '*Rotalia*' *beccarii* and *Planorbulina mediterranensis* suggest a continuing temperate climate. At Stradbroke the foraminiferal assemblages of the lower part of the Ludhamian (below +5 m OD) closely resemble those of the Ludham Crag; at the transition from pre-Ludhamian to Ludhamian an association of *Nonionella turgida* and *Textularia suttonensis* occurs which is repeated at Ludham at the Lp 1a/Lp 1b boundary within the Ludhamian, i.e. at the top of the Ludham Crag. (Beck *et al.* 1972; Funnell 1961*b*).

Of the three Ludhamian mollusc zones (Norton 1967) Lm 1 and Lm 2 correspond to the Ludham Crag, and Lm 3 to the overlying Norwich Crag horizon.

Lm 1 [= L.M.1] contains *Rissoa curticostata, Anomia squamula, Calyptaea chinensis, Hiatella arctica, Abra alba,* and *Cingula semicostata semicostata* (Norton 1967). In the lower part of the biozone-sub-zone Lm 1a—the most abundant species *Rissoa curticostata* is more dominant, and extinct species less abundantly represented, than in the upper part—sub-zone Lm 1b.

Lm 2 [= L.M.2] contains *Abra alba, Calyptraea chinensis, Caecum glabrum, Rissoa curticostata, Anomia squamula, Turritella triplicata triplicata, Mytilus edulis, Chrysallida spiralis,* and *Hiatella arctica* (Norton 1967).

Both biozones occur in the lower part of the Ludhamian (pollen biozone Lp 1a), with Lm 2 apparently extending slightly into the base of the upper part of the Ludhamian, pollen zone Lp 1b. Lm 3 also occurs in the upper part of the Ludhamian.

Lm 1 and Lm 2 may correlate with the *Mya arenaria–Hydrobia ulvae* assemblage zone (= MOL.A) of the Netherlands zonation (Zagwijn 1974; Spaink 1975).

As with the foraminifers, the Ludham Crag molluscs include a significant number of species known from the Red Crag but not from the Norwich Crag of surface outcrops. These include *Caecum glabrum,* etc. On the other hand *Serripes groenlandicus* and *Macoma calcarea,* typical of the later Red Crag, and several species characteristic of the Norwich Crag are already present. Water depths of between 40 and 15 m, shallowing upwards, are inferred. In the upper part of the Ludhamian (corresponding to uppermost Lp 1a and Lp 1b) pre-Norwich Crag forms are almost entirely absent, again corresponding with the situation amongst the foraminifers. Water depths around 15 m are inferred, with the intertidal and brackish water species *Hydrobia ulvae* appearing for the first time. The molluscs from Stradbroke have not yet been examined.

Thurnian Stage

The Thurnian Stage (West 1961) occurs directly above the Ludhamian at Ludham. Cold, subarctic park conditions are indicated by the pollen. The foraminifers (Lf 4) are reduced and consistent with colder conditions whilst the molluscs disappear almost completely in this clayey interval. This stage may also have been sampled between

−26 and −43 m OD at Southwold (West and Norton 1974), but the foraminifers from those deposits have not yet been examined and there are no molluscs. In the Ludham borehole there may well be a non-sequence or disconformity within the Thurnian (Funnell 1961*b*).

Antian Stage

The succeeding Antian Stage (West 1961) shows a return to temperate conditions and full restoration of foraminiferal and molluscan faunas of Norwich Crag type. Antian deposits have been recognized in the lower part of the section at Easton Bavents (Funnell and West 1962) and from a section at Aldeby (Norton and Beck 1972). It was originally suggested, on the basis of the foraminifers (Funnell 1961*b*), that the lower part of the type Norwich Crag section at Bramerton, near Norwich, belongs to this stage. This correlation is not contradicted by the molluscs (Norton 1967), and pollen from clay just above the stone bed at Blake's pit, Bramerton has been interpreted as Thurnian (Beck 1971).

Baventian Stage

The Baventian Stage (West 1961; Funnell and West 1962), with pollen evidence for cold conditions, is rather more widely known, both from Ludham and the type locality of Easton Bavents, from Aldeby, Norfolk (Norton and Beck 1972), between −7 and −9 m OD, at Sizewell, Suffolk (West and Norton 1974), and also from the 'Weybourne Crag' of West Runton, Norfolk (West and Wilson 1966). Funnell (1961*b*) suggested the equivalence of the uppermost beds of the Norwich Crag at Bramerton with an outcrop of 'Weybourne Crag' at Sidestrand on the basis of the foraminifers. All these sites (with the exception of Aldeby and Sizewell, from which the foraminifers have not yet been examined) are characterized by faunas lacking '*Rotalia*' *beccarii* but having *Elphidium orbiculare* and other boreo-arctic species such as *Elphidium barletti*, *Elphidium subarcticum*, and *Elphidiella groenlandicum*.

Molluscs are generally absent from the clayey acme facies of the Baventian at Ludham and Easton Bavents. In the early part of the stage at Ludham they are similar to, or a slightly reduced version of, the Antian assemblages. At Sidestrand they include the distinctive and abundant *Macoma balthica*. This species is unknown from all other localities attributed above to the Baventian stage, but is commonplace in the 'Weybourne Crag' and later marine deposits on the north Norfolk coast and elsewhere. This in itself may be a reason for questioning the attribution of 'Weybourne Crag' horizons to the true Baventian.

'Chillesford' pollen spectra

Following the Baventian of East Suffolk is a sequence of deposits including the *Scrobicularia* Crag, the Chillesford Crag and Chillesford Clay, and their possible lateral equivalent, Harmer's 'lower' division of the Norwich Crag. These have recently been shown by West and Norton (1974) to contain characteristic 'Chillesford' pollen spectra of temperate aspect, but lacking the frequencies of *Tsuga* found in the earlier Ludhamian and Antian temperate stages. Analysis of the molluscs from these deposits has confirmed the very shallow-water, often inter-tidal, conditions under which they accumulated (see Funnell 1961*b*). The pollen suggests equation with the Pastonian (West and Wilson 1966) of the north Norfolk coast, rather than to the Antian tentatively suggested by Funnell on the basis of the foraminifers. West and Norton (1974) allocate the extensive beach plain gravels of the Westleton Beds of east Suffolk and Norfolk (Hey 1967) to an episode slightly later than the 'Chillesford' phase.

Pastonian Stage

The Pastonian Stage (West and Wilson 1966) includes marine and freshwater sediments exposed and at shallow depth on the north-east Norfolk coast. At Paston, on the foreshore, marine sands and clays contain a temperate pollen flora (Pa II, III) and rest on marine sands and clays with a pollen flora with high NAP, Gramineae, and Ericales (pre-Pastonian ? = Baventian). The most complete sequence of the temperate stage has been recorded at Beeston Cliffs, where a freshwater mud and fluviatile gravels show a *Pinus* biozone (Pa I), and are overlain by tidal clays and beach gravel (Pa II (*Pinus, Quercus, Ulmus, Carpinus*) and III (?) (*Pinus, Betula, Quercus*)) and terrestrial woodpeat (Pa IV (*Pinus, Picea, Betula*, Ericales)), the whole representing an interglacial type of cycle, with a middle marine transgression which deposited tidal clays at West Runton (*West Runton Clay*), Overstrand, Sidestrand, Mundesley, Paston, Bacton, Happisburgh, Corton, and Pakefield. The faunas associated with this transgression include abundant *Macoma balthica*. The problem of correlation of the Pastonian temperate biozones with that of Chillesford has been discussed earlier.

Beestonian Stage

The Beestonian Stage (West and Wilson 1966) includes freshwater sediments with a cold stage (pollen and macroscopic remains exposed in the cliff sections at Beeston and Woman Hithe e.g. *Woman Hithe Clay* at West Runton), the overlying beach gravels with occasional shells (Woman Hithe Gravels) in the same areas, and a freshwater marl at the base of the *West Runton Freshwater Bed* at West Runton. The last contains a 'late-glacial' pollen flora which precedes the *Pinus–Betula* zone (Cr I) of the Cromerian stage. Permafrost features are associated with the lower of the freshwater beds. There is little information about the fauna and flora of the beach gravel, but it is bounded by freshwater sediments with a cold flora.

Cromerian Stage

The Cromerian Stage (West and Wilson 1966) includes marine and freshwater sediments exposed at the foot of the cliffs on the north-east Norfolk and east Suffolk coasts. The most complete sequence, displaying a full interglacial type of cycle, is seen at West Runton, where the *West Runton Freshwater Bed* (Reid's 'Upper Freshwater Bed') shows pollen biozones Cr I (*Pinus, Betula*) and Cr II (*Pinus, Quercus, Ulmus, Tilia*). A marine transgression deposited the *Yoldia [Leda] myalis Bed* in Cr III (*Carpinus, Abies*) and a regression occurred later, giving freshwater sediments in Cr IV (*Betula*). The Cromerian biozones are recorded in sediments round the Norfolk coast from West Runton to Ostend, and in Suffolk from Hopton to Kessingland. They thus occupy a wide and shallow basin between Sheringham and Covehithe. The freshwater sediments of Cr II are recorded as organic mud or alluvium (*Corton Rootlet Bed*) at Side-strand, Mundesley, Hopton, Corton, Pakefield, and Kessingland. The middle marine horizon (*Yoldia [Leda] myalis Bed, Mundesley Clay*) of Cr III is recorded at West Runton, Overstrand, Sidestrand, Trimingham, Mundesley, Paston, Bacton, Corton, and Pakefield. The freshwater sediments of Cr IV are found at West Runton, Overstrand, Mundesley and Ostend.

Anglian Stage

The glacial Anglian Stage (Mitchell *et al.* 1973) includes freshwater sediments with a cold flora (*Arctic Freshwater Bed*) overlying Cromerian sediments and underlying the first Anglian till of the coast sections. It has been recorded at Mundesley and Ostend, and is associated with permafrost features.

References

Beck, R.B. (1971). The Lower Pleistocene geology and vegetational history of East Anglia. Thesis, University of Cambridge.

Beck, R.B., Funnell, B.M. and Lord, A.R. (1972). Correlation of Lower Pleistocene Crag at depth in Suffolk. *Geol. Mag.* 109, 137–39.

Funnell, B.M. (1961a). The climatic and stratigraphic significance of the early Pleistocene foraminifera of the North Sea Basin. Ph.D. thesis, University of Cambridge.

Funnell, B.M. (1961*b*). The Paleogene and Early Pleistocene of Norfolk. *Trans. Norfolk Norwich Nat. Soc.* 19, 340-56.

Funnell, B.M. and West, R.G. (1962). The Early Pleistocene of Easton Bavents, Suffolk. *Q. Jl. geol. Soc. Lond.* 117, 125-41.

George, T.N. (Chairman) (1969). Recommendations on stratigraphical usage. *Proc. geol. Soc.* 1656, 139-66.

Harmer, F.W. (1900). The Pliocene deposits of the East of England. II, The Crag of Essex (Waltonian) and its relation to that of Norfolk and Suffolk. *Q. Jl. geol. Soc. Lond.* 56, 705-38.

Harmer, F.W. (1914-1925). The Pliocene Mollusca of Great Britain. Vols 1-2. Palaeontographical Society Monograph, London.

Hey, R.W. (1967). The Westleton Beds reconsidered. *Proc. Geol. Ass.* 78, 427-45.

Hey, R.W. (1976). Provenance of far-travelled pebbles in the pre-Anglian Pleistocene of East Anglia. *Proc. Geol. Ass.* 87, 69-81.

Mitchell, G.F., Penny, L.F., Shotton, F.W. and West, R.G. (1973). A correlation of Quaternary deposits in the British Isles. *Geol. Soc., Lond., Special Report*, 4, 1-99.

van Montfrans, H.M. (1971). *Palaeomagnetic dating in the North Sea basin.* Princo N.V., Rotterdam.

Norton, P.E.P. (1967). Marine molluscan assemblages in the Early Pleistocene of Sidestrand, Bramerton and the Royal Society Borehole at Ludham, Norfolk. *Phil. Trans. R. Soc.* B253, 161-200.

Norton, P.E.P. (1970). The Crag Mollusca—a conspectus. *Bull. Soc. belge Géol. Paléont. Hydrol.* 79, 157-66.

Norton, P.E.P. and Beck, R.B. (1972). Lower Pleistocene molluscan assemblages and pollen from the Crag of Aldeby (Norfolk) and Easton Bavents (Suffolk). *Bull. geol. Soc. Norfolk.* 22, 11-31.

Prestwich, J. (1871). On the structure of the Crag Beds of Norfolk and Suffolk. *Q. Jl. geol. Soc. Lond.* 27, 115, 325, 452.

Reid, C. (1882). The Geology of the country around Cromer. *Mem. geol. Surv. U.K.*

Spaink, G. (1975). Zonering van het mariene Onder-Pleistoceen en Plioceen op grond van mollusken fauna's. In *Toelichting bij Geologische Overzichtskaarten van Nederland.* (ed. W.A. Zagwijn and C.J. van Staalduinen), pp. 118-22. Rijks Geologische Dienst, Haarlem.

West, R.G. (1961). Vegetational history of the Early Pleistocene of the Royal Society borehole at Ludham, Norfolk. *Proc. R. Soc.* B 155, 437-53.

West, R.G. and Norton, P.E.P. (1974). The Icenian Crag of southeast Suffolk. *Phil. Trans. R. Soc.* B268, 1-28.

West, R.G. and Wilson, D.G. (1966). Cromer Forest Bed Series. *Nature, Lond.* 209, 497-98.

Wood, S.V. (1848-1882). *A monograph of the Crag Mollusca, Parts I-IV, Supplements 1-3.* Palaeontographical Society Monograph, London.

Wood, S.V. Snr. (1886). On the structure of the Red Crag. Explanation of the diagram-section by S.V. Wood Jnr. *Q. Jl. geol. Soc. Lond.* 22, 538.

Woodward, H.B. (1881). The Geology of the country around Norwich. *Mem. geol. Surv. U.K.*

Zagwijn, W.H. (1974). The Pliocene-Pleistocene boundary in western and southern Europe. *Boreas* 3, 75-97.

19

Glacial–interglacial stratigraphy of the Quaternary in Midland and eastern England

F. W. SHOTTON, P. H. BANHAM, and W. W. BISHOP

Abstract

Eastern England, from Yorkshire to Essex, and the western part of the English Midlands are chosen for critical examination. In both regions Devensian deposits may be studied close to the limit of the latest ice-sheet whilst south of this limit there is good preservation of deposits of the preceding glaciations and interglacials. In both areas there is clear evidence of three major and often complex periods of glaciation between which the Ipswichian and Hoxnian Interglacials are intercalated. Evidence is produced to support the argument that the Wolston sequence is post-Hoxnian and is therefore not the correlative of the Lowestoft Till (Anglian). It can therefore be retained as the stratotype of the penultimate glaciation.

Introduction

We are concerned primarily with sequences exposed in Midland England and East Anglia that consist of superimposed formations representing appreciable spans of Quaternary time (for localities mentioned in text, see Fig. 19.1). These have provided the evidence upon which a British Quaternary stratigraphic nomenclature has been based (Mitchell *et al.* 1973). In emphasizing these two regional successions we apologize for giving scant recognition to work carried out in northern England, Wales, and Scotland. However, although detailed late-glacial and postglacial sequences have been established for these areas, there are few exposed Ipswichian or older deposits.

In the remaining areas of southern and south-west England conflicting claims have been made concerning the existence or absence of vestigial remnants of till sheets recording the extent of former glaciation. Localized outcrops of solifluction, loess, and terrace deposits occur but no extended history of glaciation can be discerned.

East Anglia

A survey of the glacial stratigraphy of East Anglia (Turner 1973) reveals three broad divisions. The lowest (Anglian) and the highest (Devensian) appear relatively straightforward; the intermediate 'complex' (Hoxnian–Wolstonian–Ipswichian) requires some discussion. These divisions will be dealt with in stratigraphical order, with an emphasis on the results of recent work.

Anglian

The lithostratigraphy of the type site of this stage is at Corton (see Banham 1971). Here the chalky Lowestoft Till overlies a sandy Cromer Till; two additional Cromer Tills are found farther to the north in Norfolk (Fig. 19.2). The three Cromer Tills are remarkably similar, uniform, well-sorted, silty to sandy tills (Perrin *et al.* 1973). Both First and Second Cromer Tills are compact, thin and foliated and are probably lodgement tills; the Second is distinctively chalky; the Third Cromer Till is much thicker and contains rafts of sands; it is possibly a melt-out till in large part. Their structures (Banham 1975), mainly rolled flint erratics, marine-type sand grains (Krinsley and Funnell 1965), and clay minerals (Kazi 1972), all suggest derivation from the floor of the North Sea to the north-east.

The sands between the Cromer Tills are generally cross-bedded, and apparently laid down in shallow water under variable current conditions. Turbidite units have been recognized in the Intermediate Beds at Happisburgh (Kazi and Knill 1969). Ice-wedge casts in the Corton and Mundesley Sands and the absence of an indigenous flora

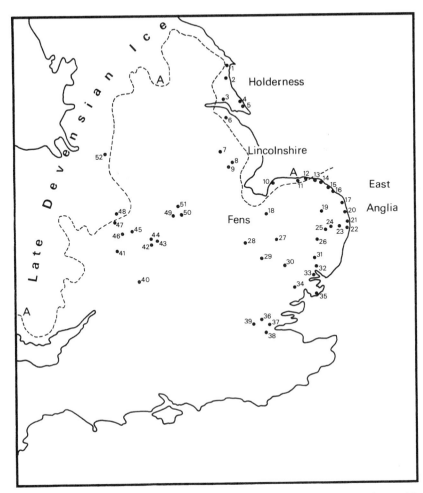

Fig. 19.1. Locality map. AA—limit of Late Devensian ice; Aveley—37; Baginton—42; Barham—31; Blakeney—12; Broome—23; Brundon—30; Cambridge—29; Chelford—52; Clacton—35; Corton—21; Cromer—14; Coventry—44; Dimlington—4; Earith—28; Four Ashes—48; Gimingham—15; Happisburgh—17; Hessle—3; Homersfield—24; Hornchurch—6; Hoxne—26; Hunstanton—10; Ilford—39; Ipswich—32; Kirkby on Bain—8; Kirmington—6; Leicester—49; Lowestoft—22; Marks Tey—34; Mildenhall—27; Moreton-in-Marsh—40; Morston—11; Mundesley—16; Nechells—45; Norwich—19; Oadby—50; Quinton—46; Scratby—20; Sewerby—1; Sheringham—13; Skipsea—2; Stutton—33; Swanscombe—38; Syston—51; Tattershall—9; Upton Warren—41; Withernsea—5; Wolston—43; Wolverhampton—47; Wortwell—25; Wragby—7; Wretton—18.

and fauna (West and Wilson 1968) indicate cold conditions throughout.

The sandy tills of the Norwich Brickearth of inland Norfolk are lithologically similar to the Cromer Tills (Perrin *et al.* 1973) and are

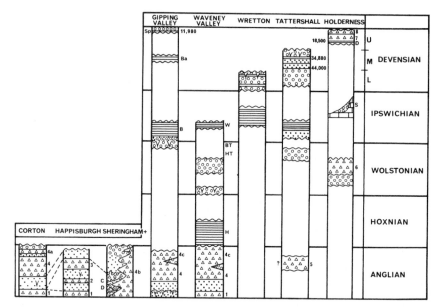

Fig. 19.2. Correlation of glacial and interglacial deposits between sections in eastern England. Only approximate thicknesses and lateral equivalences are indicated within stages. Key: Tills: 1, 2, 3—First, Second, and Third Cromer Tills; 4—Lowestoft Till; 4a—Pleasure Gardens Till; 4b—Marly Drift; 4c—Gipping Till; 5—Wragby Till; 6—Basement Till; 7—Drab (Skipsea) Till; 8—Purple (Withernsea) Till; CD—Contorted Drift. Other symbols: B—Bobbits Hole lake muds etc.; Ba—Barham Terrace deposits; BT—Broome Terrace; HT—Homersfield Terrace; D—Dimlington Silts; H—Hoxne lake muds etc.; S—Sewerby raised beach; Sp—Sproughton; W—Wortwell lake muds etc.; 44 000 (etc)—radiocarbon dates; frost-wedges are indicated. There is no evidence at Tattershall itself for placing the chalky till at the base of the section in the Anglian rather than in the Wolstonian.

generally considered to be their lateral equivalents.

South of Scratby the Cromer Tills are overlain by the Lowestoft Till. This extensive till is characterized by a clay-rich matrix and a high proportion of chalk erratics. Preferred orientations of erratics appear to indicate ice-movement from the west (West and Donner 1956).

Elsewhere, the relations between the Cromer and Lowestoft Tills are more complex. In the Norwich area (Cox and Nickless 1972) interdigitation occurs; to the north of Norwich, Cromer and Lowestoft Tills are separated laterally by a northward widening outcrop of Marly Drift (Straw 1965). Recent work by Perrin and Davies (in Banham *et al.* 1975) on the mechanical and mineralogical composition of this deformed and chalk-enriched till complex has demonstrated two main lithological types. To the east and low in the sequence occurs the Cromer-type, whereas to the west and high in the sequence is found the Lowestoft-type of Marly Drift. It seems likely that in north Norfolk the deformation and chalk-enrichment, both of the Cromer Tills and of the Lowestoft Till itself, took place during the advance of the Lowestoft ice (Banham and Ranson 1965).

The thick sequence of Briton's Lane (Cromer Ridge) Gravels that forms the high ground near Sheringham may be outwash from this ice. In any event, the weight of these sediments appears to have induced non-directional load structures associated with diapirs in the underlying Cromer Tills (The 'Contorted Drift'; see Banham 1975).

Hoxnian–Wolstonian–Ipswichian

Fossil-bearing lake and estuary muds and marine sands overlie Anglian rocks and underlie Devensian rocks at many sites in East Anglia and provide ample evidence for interglacial climatic conditions. The evidence of the pollen, in particular, indicates two independent, full interglacial cycles, the Hoxnian and the Ipswichian (Turner 1970; Sparks and West, 1970). As by definition interglacials can be separated in time only by a glacial stage, the Wolstonian is implicit in this interpretation.

According to Turner (1970) Hoxnian floras can be distinguished from those of the Ipswichian by dominance of *Pinus* and *Alnus* throughout the temperate period, by low frequency of *Quercus* and *Corylus* even during the early temperate, by the early appearance of *Picea*, and by the abundance of *Abies* during the late temperate. The Ipswichian is further characterized by plants of Mediterranean type, especially *Pyracantha coccinea* and *Acer monspessulanum* (Phillips 1974; see also Turner in discussion of Bristow and Cox 1973).

The beetle evidence (Coope 1974 and Chapter 5, this volume) amplifies the broad conclusion that whereas the Hoxnian was a relatively cool interglacial, Ipswichian temperatures were higher than those of today.

The vertebrates, principally mammals, of Hoxnian deposits include: *Trogontherium, Pitymus, Ursus, Arvicola,* and *Lemmus,* whereas the Ipswichian is characterized by *Emys, Mammuthus, Crocuta,* and *Hippopotamus.* In view of the climatic cycles envisaged, it is not surprising that several of these genera are also found in deposits of other ages, and that yet others are found in both Hoxnian and Ipswichian sequences (Stuart 1974; see also Gibbard and Stuart 1975).

Further support for the recognition of two distinct interglacials is provided by fluvial geomorphology. Hoxnian freshwater sites occur on modern interfluves in the main, and can be related to sea-levels up to 40 m OD as indicated by shore deposits. Ipswichian sites, by contrast, occur at low levels, often in terrace deposits related to modern streams, and shore deposits have not been found above 15 m OD (West 1972).

Wymer (1974 and Chapter 8, this volume) has concluded from a review of the archaeological evidence that no clear distinction between Hoxnian and Ipswichian sequences can be made on artifact typology alone. However, a consistent pattern of changes can be seen in relation to the pollen spectra. Whereas Hoxnian sequences contain Clactonian and Acheulian industries, the Ipswichian is characterized by the Acheulian and Levalloisian.

Finally, provisional uranium method dating of bones (Szabo and Collins 1975) has shown large, if variable, time gaps between probable Hoxnian and later material:

Hoxnian	Ipswichian/Wolstonian
Swanscombe—over 272 000	Stutton—125 000 ± 20 000
Clacton—245 000 $^{+\,35\,000}_{-\,25\,000}$	Brundon—174 000 ± 30 000
(See also Suggate 1974)	

Taken together, these various lines of evidence strongly support the separation by a glacial stage of an older, cooler, Hoxnian from a younger, warmer Ipswichian Interglacial stage. Nowhere outside a stratigraphical table is an Ipswichian sequence known to occur vertically above a Hoxnian sequence, however. Further, although cold-indicating fossils and cryoturbated sediments not uncommonly overlie and underlie Hoxnian and Ipswichian rocks, respectively, no Hoxnian section is capped by independently dated pre-Devensian beds, and no Ipswichian sequence rests upon undoubted post-Anglian deposits.

In the Gipping valley, especially, can be seen a sequence of dark grey, chalky, Lowestoft Till overlain by barren sands and gravels, overlain in turn by a creamy-brown chalky Gipping Till. Early studies of petrology and pebble orientations led to the conclusions that this till could be distinguished from the Lowestoft Till below and, moreover, that it was younger than Hoxnian deposits. More recently, detailed mapping and borehole interpretation (Straw 1965; Turner 1970; Woodland 1970; and, especially, Bristow and Cox 1973) and lithological analysis (Perrin *et al.* 1973) all strongly indicate that the Gipping Till is better regarded as a weathered component of the Anglian Lowestoft Till.

At High Lodge, Mildenhall, two chalky tills and their associated gravels, separated by tool- and fossil-bearing 'interstadial' lake muds, have been assigned to the Wolstonian (Turner 1973; Wymer 1974). However, preliminary lithological work by Perrin *et al.* (1973) on the Breckland Tills has revealed the presence of a low-clay, high-zircon possibly re-worked variation on the theme of the uniform and ubiquitous Lowestoft Till. The chalky till of the nearby Cambridge district has been correlated with the Lowestoft Till (Worssam and Taylor 1969), although an apparently similar till in the Ouse and Nene basin (Horton 1970) has not been designated, perhaps wisely, in view of the stratigraphical difficulties involved (see Fig. 19.3).

Other supposedly Wolstonian deposits, such as the near-surface sands and gravels at Blakeney, can be demonstrated only as post-Marly Drift (Anglian) and pre-Hunstanton Till (Devensian). As Straw (1973) pointedly notes, a remarkably large quantity of out-wash exists with no evidence of a related till in this interpretation.

From the scant evidence available, it appears that the Wolstonian was climatically a glacial period during which no till was deposited in East Anglia. Sea-levels were generally low, however, causing major valley incision followed by partial infilling by sands and gravels which may be re-worked Anglian deposits in large part. Long, cold periods with no till deposition are of course well-known in the Devensian (Coope 1975a); nevertheless, the existence of the thick, multi-till Wolstonian sequences of the Midlands should be a major stimulus for further investigations in both areas.

Devensian

Devensian deposits are best preserved in the low Fenland where they are found in the terrace deposits of modern streams. At Wretton (West *et al.* 1974) Ipswichian deposits are overlain by sands and gravels containing lenses of organic silts. Interpretations of the frost structures, pollen, beetles, and molluscs concur in the general conclusion that cold, tundra conditions prevailed. These early Devensian rocks are beyond the range of carbon-14 techniques.

A somewhat similar, younger (42 000 BP, Middle Devensian) sequence at Earith (Bell 1970) contains a dwarf willow flora once again indicating cold, high steppe conditions. A slightly younger ($37\,720\,^{+1670}_{-1390}$ BP) fossiliferous sequence of even colder aspect is known from low terrace deposits at Syston Leicestershire (Bell *et al.* 1972).

Perhaps the most remarkable sections in the Devensian rocks of the Fens occur at Tattershall and Kirkby-on-Bain in Lincolnshire. Preliminary descriptions by Girling (1974) reveal a basal, chalky till, the Wragby Till of Straw (1966), and gravels unconformably overlain by a shell bed containing land molluscs, and then by a peat bed. From the pollen this peat has been dated as Ipswichian (Ip IIb; Evans and Phillips quoted in Girling 1974). This in turn is unconformably overlain by gravels and sands containing one of the richest mammal

faunas in Britain, including bison, reindeer, mammoth, woolly rhinoceros, red deer, *Megaceros*, horse, and bear.

Two silt beds occur in these gravels at Tattershall. The lower
(44 000 BP; see Shotton *et al.* 1973*a, b,* 1974 for all dates), contains
a cold insect fauna, whereas the upper (Anodonta Bed), which is
probably only 1000 years younger, contains a fauna indicative of the
peak warmth of the Upper Warren interstadial (Coope 1975*b*). A
somewhat younger silt at Kirkby (34 800 ± 1000 BP) once again has
a cool fauna. Although the overlying gravels cannot be dated, they
contain massive ice-wedge pseudomorphs which clearly demonstrate
the recurrence of cold conditions. (See also Sparks *et al.* 1972;
Williams 1975).

Thin but widespread, near-surface, windblown deposits have been
reported from Norfolk, Lincolnshire, and Yorkshire (Catt *et al.*
1971, 1974; Perrin *et al.*, 1974). These deposits are mineralogically
similar to the Devensian tills which in places rest directly upon them,
and are thought to have been picked up by wind from outwash
deposited in advance of the ice. (See also Chapter 16.)

The extent of the narrow, coastal strip of Devensian tills in eastern
England is well known (Suggate and West 1959; Straw 1960).
Recent, refined mineralogical work by Madgett (1975) has shown
that the upper Hessle Till of Holderness is merely the weathered zone
of the underlying Purple (Withernsea) and Drab (Skipsea) Tills. The
latter is the thicker and more extensive, and has been traced via the
Marsh Till of Lincolnshire into the Hunstanton Till of west Norfolk.
That these tills were emplaced during approximately 5000 years
following the deposition of the Dimlington Silts (18 500 BP) has
been shown by Penny *et al.* (1969). Bell and Dickson (1971) report a
cold flora and fauna from the Barnwell Station Beds which were
deposited at about this time (19 500 BP) in a channel now buried
beneath the river Cam at Cambridge.

The evidence from the Devensian of East Anglia is thus in accord
with the broadly based conclusion of Coope (1975*a*) that a long
period of cold, arid conditions, interrupted by the (Wretton), Chelford, and Upton Warren Interstadials, preceded the arrival of ice
from the North Sea towards the end of the stage; the subsequent
climatic amelioration has been thoroughly documented by Walker
and West (1970).

Midland England

Devensian

The Pleistocene chronology of Midland England is best understood
by working back through the time succession. The Cheshire low-

lands are covered by Quaternary deposits resulting from the south-eastward advance of a glacier from the Irish Sea basin. Thin sands usually overlie a till or a glacio-lacustrine clay ('Upper Boulder Clay') which in turn lies upon sand with occasional intercalations of till (so-called 'Middle Sand'). These constitute the Stockport Formation. Marine shells occurring in the pre-till sand, radiocarbon dated at 28 000 ± 1800 BP show that the till is Late Devensian. In places a much older sand, brown to white in colour, is covered unconformably by the 'Middle Sand', and has been called the Congleton Sand. In two places it incorporates a thick peat bed with trunks of *Pinus, Picea,* and *Betula.* Its flora at Chelford (Simpson and West 1958) indicated a correlation with the Brørup Interstadial of Denmark and so does an 'enhanced' radiocarbon date of 60 800 BP (GrN –1475). An extensive insect fauna has also been described (Coope 1959).

Undoubtedly a complex of older Pleistocene deposits occurs beneath the Cheshire plain, but it has not been interpreted. The Upper Till can, however, be traced southwards to end in a broad, low moraine trending roughly east–west through Wolverhampton. A few kilometres north of this, a pit at Four Ashes (Morgan, A.V. 1973) provided the following section which has been taken as the stratotype for the last, Devensian, glacial stage:

(4)   Sands;
(3)   Irish Sea Till;
(2)   Sands and gravels with silt lenses;
(1)   Relic peat with an interglacial flora (Ipswichian), resting on Trias.

Peats which rest upon sand (4) further north have been dated around 13 500 BP so that the large and complicated ice-wedge pseudomorphs which cut all the Four Ashes succession are earlier than this but post-date the till.

The gravels under the till cover the Middle and Lower Devensian. There were many silt lenses whose insect fauna was studied by Anne Morgan (1973) and seven of these were radiocarbon dated. The youngest dates, which accompany a very cold fauna, prove that the till is more recent than 30 500 ± 400 BP. A much warmer fauna accompanied a date of 42 530 BP and clearly represents the Upton Warren Interstadial (see below). Beyond the reach of radiocarbon dating were found both cold and warmer faunas, and the latter was suggested to be equivalent to the Chelford Interstadial. Palynology has since confirmed this suggestion.

South of the Wolverhampton moraine, Devensian deposits are found only in the river terraces, and older glacial sediments are left as erosional remnants on the interfluves or filling channels of a highly anomalous ancient drainage system. No. 2 terrace of the Avon (Tom-linson 1925) has a cold, even arctic, flora and fauna and several radiocarbon dates between 28 000 and 38 000 BP. The Main Terrace of the Severn (Wills 1938), on its tributary the River Salwarpe, has given dates of 42 000 BP and yielded a warmer, continental fauna which has defined the warm peak of the Upton Warren Interstadial Complex (Coope *et al.* 1961). There are higher, older terraces in the Severn, the Avon, and the north-flowing Tame, and in one of these on the Avon, *Hippopotamus* and the molluscs *Corbicula fluminalis* and *Belgrandia marginata* indicate the Ipswichian Interglacial.

Wolstonian

Shotton (1953, 1968) working south-east of Coventry mapped a stratified Pleistocene succession preserved on the interfluves between valleys of the River Avon and its tributaries. This has been chosen as the type for the Wolstonian glacial stage. The sequence is demon-strably pre-Ipswichian as river terraces of Ipswichian and Devensian age are incised below its deposits. The fauna of its basal member (in-cluding *Mammuthus primigenius, Palaeoloxodon antiquus, Equus caballus, Rangifer tarandus, Coelodonta antiquitatis, Crocuta cro-cuta,* and *Dicrostonyx torquatus*) strongly suggests that it is not older than the Saalian of Europe. Following the work of Rice (1968) in the Leicester area, Shotton (1976) has redefined the units within the Wolstonian as follows:

Dunsmore Gravel
Upper Oadby Till
Lower Oadby Till   Previously grouped as
Wolston Sand   Upper Wolston Clay,
Bosworth Silt and Clay   Wolston Sand, and Lower
Thrussington Till   Wolston Clay.
Baginton Sand
Baginton–Lillington Gravel

Between the Thrussington Till and Lower Oadby Till is a retreat of interstadial proportions with the Bosworth Silt and Clay laid down in the large ice-impounded Lake Harrison (Shotton 1953).

In the Coventry–Leicester area only small patches of an earlier till (Bubbenhall Clay) are preserved under the Baginton beds and these are referred to the Anglian stage. No Hoxnian deposits have yet been discovered here. To the west, however, from Coleshill to Birmingham (all localities lying south 'of the Wolverhampton Moraine) several Hoxnian sequences are known, all in positions unrelated to the present drainage pattern. The two most significant are in north Bir-mingham at Nechells (Kelly 1964, Shotton and Osborne 1965) and in West Birmingham at Quinton (Horton 1974). At both these sites silts, clays and peats fill a hollow cut into an underlying glacial sequence and are covered by a subsequent glacial series. The pollen stratigraphy matches Hoxne and Marks Tey zone by zone. Discussion continues on how the two sequences, 10 km apart, should be connected on a map, but there is no dispute on the basic facts that at Quinton a thick till lies beneath the interglacial deposit and another lies above it, and at Nechells there are gravels of fluvio-glacial type beneath the interglacial deposit and a multiple sequence of tills and gravels above it.

There is much more supporting evidence in the West Midlands, but space demands prevent it being cited. Comment has been restricted to the minimum of fact necessary to support the correlation scheme of Fig. 19.3.

Correlation

Evidence for only one major lithological unit of Chalky Till occurs in each of the two regions described above. Considerable discussion of possible correlations has ensued (see Mitchell *et al.* 1973, p. 9).

In East Anglia and southward *into* the Thames valley a sheet of Chalky Tills is Anglian by definition as it includes the type section. This sheet is overlain by Hoxnian interglacial deposits at Hoxne and Marks Tey. Additionally, terrace deposits of the Thames and its tributaries, comprising the Boyn Hill/Swanscombe/Clacton complex, which is equated with the Hoxnian, are younger than the sheet of Chalky Tills. This is based upon evidence at Hornchurch in Essex where Boyn Hill terrace gravels of the proto-Thames overlie a lobe of Chalky Till in a position inferred to be close to the original southern limit of the ice-sheet. Nowhere in Suffolk or Essex has a second Chalky Till been found resting upon the Hoxnian.

In the Birmingham and Wolverhampton areas of the West Midlands there is evidence of three glacials and two interglacials. The Quinton and Nechells successions each show two glacial sequences separated by Hoxnian interglacial deposits. Geomorphologically these successions clearly antedate flights of terraces within the Severn–Avon drainage system and the Devensian succession of Four Ashes, which rests upon an Ipswichian deposit.

The Severn–Avon river terraces contain palaeontological evidence which indicates that they span the Ipswichian and Devensian but no evidence of any older deposits has been found. Terraces at Upton Warren and other localities provide good evidence of interstadial conditions within the Devensian. To the north-west in the Wolverhampton area, Irish Sea Till is underlain by Ipswichian deposits succeeded by gravels containing silty lenses with plant material yielding radiocarbon ages indicating the Upton Warren interstadial.

The till is overlain by peaty deposits yielding late-glacial pollen and radiocarbon dates. These Devensian glacial and interstadial deposits are the lateral equivalent of the Devensian terrace sequence which occurs a few miles to the south and west along the valleys of the River Avon and Severn.

There is some division of opinion about the position of the Wolston succession which has been selected as the stratotype for the glaciation between the Hoxnian and Ipswichian. Arguments have been advanced by some workers, based solely upon lithological similarity, that the Oadby Till of Leicestershire and Warwickshire and the Chalky Till of East Anglia should be equated and assigned to a single, if complex, stage of glaciation before the Hoxnian (Bristow and Cox 1973; Perrin *et al.* 1973). Since the Anglian Chalky Till(s) lies demonstrably between the Cromerian and the Hoxnian, it would then follow that the whole of the Wolstonian must be pre-Hoxnian, since the Chalky Till of Oadby is the newest member of the sequence.

Against this view is the purely negative one that the lithology of a till depends upon the rocks which are being eroded and similarity of lithology need not imply synchroneity; and the rather less negative point that relics of an older glaciation are known beneath the Wolstonian (the Bubbenhall Clay), even though no Hoxnian has yet been found in the type area. Farther south, at Moreton-in-Marsh, Wolstonian Till at its limit of extension has below it the Paxford Gravel with *Mammuthus primigenius* and below this the Stretton Sand which has yielded a small fauna including *Palaeoloxodon antiquus*, adjudged to be Hoxnian, not Cromerian (Shotton 1973). In the Avon terraces postdating the Wolstonian, the Ipswichian is clearly represented by deposits containing a *Hippopotamus–Corbicula* fauna, but no Hoxnian terrace is known. Finally, the mammal fauna from the Baginton–Lillington Gravel at the base of the Wolstonian (see p. 126), could very satisfactorily be post-Hoxnian but seems to be incompatible with a late-Cromerian or earliest Anglian age which it would have to be if the Chalky Tills of Leicestershire were correlated with the Chalky Anglian Tills of the type area.

In Fig. 19.3, therefore, we set out the most simple interpretation of the sequences in the two regions and of the correlations between them based upon local stratigraphic succession, palynology, mammalian fossils, and geomorphic relationships. The diagram implies that during three successive glaciations the ice sheets exhibited lobate margins which varied considerably in their distance of penetration southward. The offlap of progressively younger till sheets from south-east to north-west is clearly shown. During the Wolstonian the tills and glacial lake sediments of the Midlands were contemporary

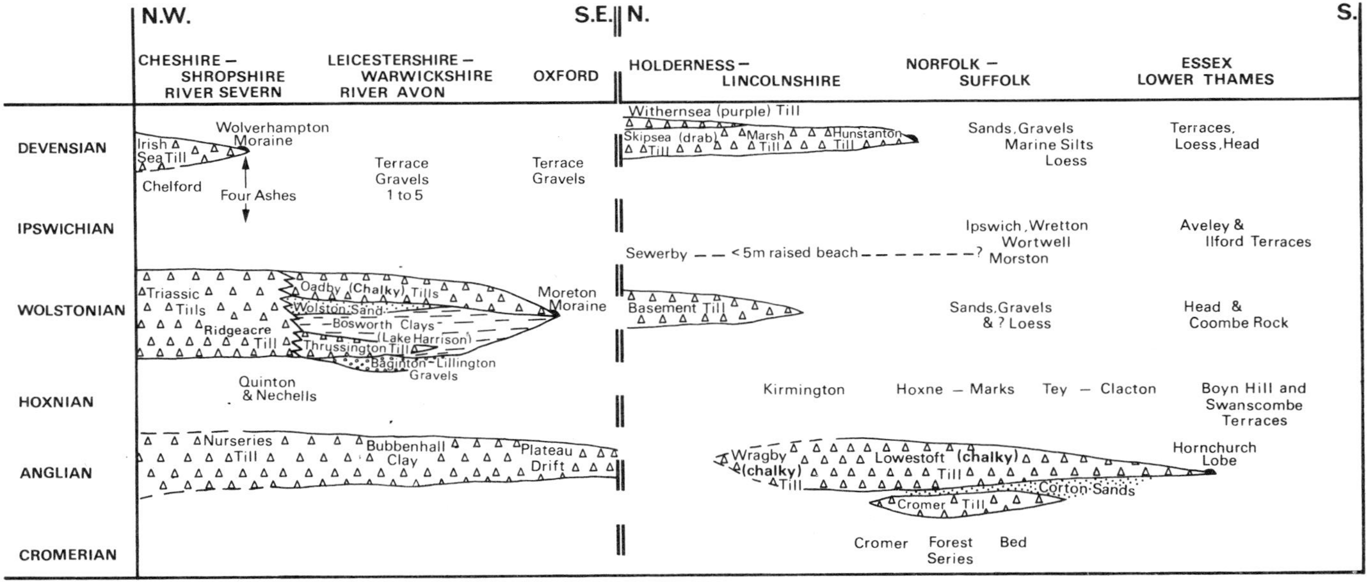

Fig. 19. 3. Correlation diagram for the Midlands (left) and eastern England (right).

280   *Glacial–Interglacial stratigraphy of the Quaternary*

with deposition of gravels in East Anglia where ice-sheets did not penetrate. This east to west contrast during the Wolstonian is similar to that described above for the West Midlands during the Devensian to the north and south of Wolverhampton.

Acknowledgements

The authors greatly benefited from discussions with others and P.H.B. is particularly indebted to J. Rose, J.A. Catt, C.P. Green, and N. Sinclair-Jones.

References

Banham, P.H. (1971). Pleistocene beds at Corton, Suffolk, *Geol. Mag.* 108 (4).
Banham, P.H. (1975). Glacitectonic structures: a general discussion with particular reference to the Contorted Drift of Norfolk. In *Ice Ages: Ancient and Modern* (ed. A.E. Wright and F. Moseley) *Geol. J. Spec. Issue* 6, Liverpool.
Banham, P.H., Davies, H., and Perrin, R.M.S. (1975). Short field meeting in north Norfolk, *Proc. Geol. Ass.* 86 (2), 251-58.
Banham, P.H., and Rawson, C.E. (1965). Structural study of the Contorted Drift and disturbed chalk at Weybourne, north Norfolk. *Geol. Mag.* 102, 164-74.
Bell, F.G. (1970). Late Pleistocene floras from Earith, Huntingdonshire, *Phil. Trans. R. Soc.* B 258, 347-78.
Bell, F.G., Coope, G.R., Rice, T.J. and Riley, T.H. (1972). Mid-Weichselian fossil-bearing deposits at Syston, Leicestershire, *Proc. Geol. Ass.* 83 (2), 197-212.
Bell, F.G., and Dickson, C.A. (1971). The Barnwell Station Arctic flora: a reappraisal of some plant identifications. *New Phytol.* 70, 627-38.
Bristow, C.R. and Cox, F.C. (1973). The Gipping Till: a reappraisal of East Anglian glacial stratigraphy. *J. geol. Soc. Lond.* 129, 1-37.
Catt, J.A., Corbett, W.M., Hodge, C.A.H., Madgett, P.A., Tatler, W. and Weir, A.H. (1971). Soils of north Norfolk. *J. Soil Sci.* 22 (4), 444-52.
Catt, J.A., Weir, A.H. and Madgett, P.A. (1974). The loess of eastern Yorkshire and Linconlshire, *Proc. Yorks. geol. Soc.* 40, 23-39.
Coope, G.R. (1959). A Late-Pleistocene insect fauna from Chelford, Cheshire. *Proc. R. Soc.* B 151, 70-86.
Coope, G.R. (1974). Interglacial Coleoptera from Bobbitshole, Ipswich, Suffolk. *J. geol. Soc. Lond.* 130, 333-40.
Coope, G.R. (1975a). Climatic fluctuations in north-west Europe since the last Interglacial, indicated by fossil assemblages of coleoptera. In *Ice Ages: ancient and modern* (ed. A.E. Wright and F. Moseley) *Geol. J. Spec. Issue* 6. Liverpool.
Cooper, G.R. (1975b). Mid-Weichselian climatic changes in Western Europe, reinterpreted from coleopteran assemblages. *Quaternary studies* (ed. R.P. Suggate and Cresswell), 101-8. R. Soc. New Zealand.
Coope, G.R., Shotton, F.W. and Strachan, I. (1961). A Late Pleistocene fauna and flora from Upton Warren, Worcestershire. *Phil. Trans. R. Soc.* B 244, 379-421.
Cox, F.C. and Nickless, E.F.P. (1972). Some aspects of the glacial history of Central Norfolk. *Bull. geol. Surv. Gt. Br.* 42.
Gibbard, P.C. and Stuart, A.J. (1975). Flora and vertebrate fauna of the Barrington Beds, *Geol. Mag.* 112 (5), 493-501.

Girling, M.A. (1974). Evidence from Lincolnshire of the age and intensity of the mid-Devensian temperate episode. *Nature, Lond.* 250, 270.

Horton, A. (1970). The drift sequence and subglacial topography of the Ouse and Nene basin, *U.K. IGS Report* 70/9, 30 pp.

Horton, A. (1974). The sequence of Pleistocene deposits proved during the construction of the Birmingham motorways. *Inst. Geol. Sci.* Report 74/11, 30 pp.

Kazi, A. (1972). Clay mineralogy of the North Sea Drift. *Nature. phys. Sci.* 240 (99), 61-2.

Kazi, A., and Knill, J.L. (1969). The sedimentation and geotechnical properties of the Cromer Till between Happisburgh and Cromer, Norfolk. *Q.J. Eng. Geol.* 2, 63-86.

Kelly, M.R. (1964). The Middle Pleistocene of North Birmingham. *Phil. Trans. R. Soc.* B 247, 533-92.

Krinsley, D.H. and Funnell, B.M. (1965). Environmental history of quartz sand grains from the Lower and Middle Pleistocene of Norfolk, England. *Q. Jl. geol. Soc. Lond.* 121, 435-61.

Madgett, P.A. (1975). Re-interpretation of Devensian Till stratigraphy of eastern England. *Nature, Lond.* 253, 105-7.

Mitchell, G.F., Penny, L.F., Shotton, F.W., and West, R.G. (1973). A correlation of Quaternary deposits in the British Isles. *Geol. Soc. Lond. Spec. Rep.* 4.

Morgan, A. (1973). Late Pleistocene environmental changes indicated by fossil insect faunas of the English Midlands. *Boreas*, 2 (4), 173-212.

Morgan, A.V. (1973). The Pleistocene geology of the area north and west of Wolverhampton, Staffordshire, England. *Phil. Trans. R. Soc.* B 265, 233-97.

Penny, L.F., Coope, G.R. and Catt, J.A. (1969). Age and insect fauna of the Dimlington Silts, East Yorkshire, *Nature, Lond.* 224, 65-7.

Perrin, R.M.S., Davies, H. and Fysh, M.D. (1973). Lithology of the Chalky Boulder Clay, *Nature phys. Sci.*, 254, 101-4.

Perrin, R.M.S., Davies, H. and Fysh, M.D. (1974). Distribution of Late Pleistocene aeolian deposits in East and South England. *Nature, Lond.* 248 (5446), 320-4.

Phillips, L.M. (1974). Vegetational history of the Ipswichian/Eemian interglacial in Britain and Continental Europe, *New Phytol.* 73, 589-604.

Rice, R.J. (1968). The Quaternary deposits of central Leicestershire, *Phil. Trans. R. Soc.* A 262, 459-509.

Shotton, F.W. (1953). Pleistocene deposits of the area between Coventry, Rugby and Leamington and their bearing on the topographic development of the Midlands. *Phil. Trans. R. Soc.* B 237, 209-60.

Shotton, F.W. (1968). The Pleistocene succession around Brandon, Warwickshire, *Phil Trans. R. Soc.* B 254, 387-400.

Shotton, F.W. (1973). A mammalian fauna from the Stretton Sand at Stretton-on-Fosse, South Warwickshire, *Geol. Mag.* 109, 473-6.

Shotton, F.W. (1976). Amplification of the Wolstonian Stage of the British Pleistocene. *Geol. Mag.* 113, 241-50.

Shotton, F.W. and Osborne, P.J. (1965). The fauna of the Hoxnian interglacial deposits of Nechells, Birmingham, *Phil. Trans. R. Soc.* B 248, 353-78.

Shotton, F.W. and Williams, R.E.G. (1973a). Birmingham University radiocarbon dates VI. *Radiocarbon* 15, 1-12.

Shotton, F.W. and Williams, R.E.G. (1973b). Birmingham University radiocarbon dates VII. *Radiocarbon* 15, 451-68.

Shotton, F.W., Williams, R.E.G. and Johnson, A.S. (1974). Birmingham University radiocarbon dates VIII. *Radiocarbon* 16, 285-305.

Simpson, I.M. and West, R.G. (1958). On the stratigraphy and palaeobotany of a late-Pleistocene organic deposit at Chelford, Cheshire, *New Phytol.* 57, 239–50.

Sparks, B.W. and West, R.G. (1970). Late Pleistocene deposits at Wretton, Norfolk. Ipswichian interglacial deposits, *Phil. Trans. R. Soc.* B 258, 1–30.

Sparks, B.W., Williams, R.B.G. and Bell, F.G. (1972). Presumed ground-ice depressions in East Anglia, *Proc. R. Soc.* A 327, 329–43.

Straw, A. (1960). Limit of the last Glaciation in north Norfolk, *Proc. Geol. Ass.* 71, 379–90.

Straw, A. (1965). A reassessment of the chalky boulder clay or marly drift of north Norfolk, *Z. fur. Geomorph.* 9 (2), 209–21.

Straw, A. (1966). The development of the middle and lower Bain valley, Lincolnshire, *Trans. Inst. Br. Geogr.* 40, 145–54.

Straw, A. (1973). The glacial geomorphology of central and north Norfolk, *E. Midld. Geogr.* 5 (7), No. 39, 333–54.

Stuart, A.J. (1974). Pleistocene history of the British vertebrate fauna, *Biol. Rev.* 49, 225–66.

Suggate, R.P. (1974). When did the last Interglacial end? *Quaternary Res.* 4 (3), 246–52.

Suggate, R.P. and West, R.G. (1959). On the extent of the Last Glaciation in eastern England, *Proc. R. Soc.* B 150, 263–83.

Szabo, B.J. and Collins, D. (1975). Ages of fossil bone from British interglacial sites. *Nature, Lond.* 254, 680–82.

Tomlinson, M.E. (1925). River Terraces of the Lower Valley of the Warwickshire Avon. *Q. Jl. geol. Soc. Lond.* 81, 137–69.

Turner, C., (1970). The Middle Pleistocene deposits at Marks Tey, Essex, *Phil. Trans. R. Soc.* B 257, 373–440.

Turner, C. (1973). Eastern England. In *Quaternary* (ed. G.F. Mitchell *et al.*), pp. 8–18. *Geol. Soc. Lond., Spec. Report* 4.

Walker, D. and West, R.G. (eds.) (1970). *Studies in the vegetational history of the British Isles*, University Press, Cambridge.

West, R.G. (1972). Relative land–sea-level changes in south-eastern England during the Pleistocene, *Phil. Trans. R. Soc.* A272, 87–98.

West, R.G., Dickson, C.A., Catt, J.A., Weir, A.H., and Sparks, B.W. (1974). Late Pleistocene deposits at Wretton, Norfolk. II. Devensian deposits. *Phil. Trans. R. Soc.* B 267, 337–420.

West, R.G. and Donner, J.J. (1956). The glaciations of East Anglia and the East Midlands: a differentiation based on stone orientation measurements of the tills, *Q. Jl. geol. Soc. Lond.* 112, 69–87.

West, R.G. and Wilson, D.G. (1968). Plant remains from the Corton Beds at Lowestoft, Suffolk. *Geol. Mag.* 105, 116–23.

Williams, R.B.G. (1975). The British climate during the Last Glaciation; an interpretation based on periglacial phenomena. In *Ice Ages: Ancient and Modern* (ed. A.E. Wright and F. Moseley), *Geol. J. Spec. Issue* 6, Liverpool.

Wills, L.J. (1938). The Pleistocene development of the Severn from Bridgnorth to the sea, *Q. Jl. geol. Soc. Lond.* 94, 161–242.

Woodland, A.W. (1970). The buried tunnel-valleys of East Anglia, *Proc. Yorks. geol. Soc.* 37, 521–78.

Worssam, B.C. and Taylor, J.H. (1969). Geology of the country around Cambridge, *Mem. geol. Surv. U.K.* 188.

Wymer, J.J. (1974). Clactonian and Acheulian industries in Britain—their chronology and significance, *Proc. geol. Ass.* 85 (3), 391–422.

20

The late Quaternary history of the climate of the British Isles

H. H. LAMB

Abstract

Although there were drastic and repeated climatic changes throughout the Quaternary, it is only towards the end of the system that the detail becomes adequate for meteorological modelling. Particular consideration is given to the last major cold period (Weichselian, Devensian, Wisconsinan, Würm) culminating in the major accretion of ice around 18 000 years ago; to the episodes of great fluctuation between 15 000 and 10 000 BP; and to the Holocene (Flandrian).

In the Weichselian, the obtaining of seasonal parameters capable of leading to reconstructions of the dominant patterns of barometric pressure and wind circulation depend largely on biological evidence and CLIMAP. In the later part of the Holocene, archaeology and then written records can contribute greatly to reconstructions of typical weather maps.

Introduction

Oxygen isotope analysis of sediment cores from the various oceans of the world has established (Shackleton and Opdyke 1976) a much greater number of cold-climate episodes in the Quaternary than were formerly known. Major events accompanied by formation of substantial ice-sheets on land are deduced to have occurred approximately every 100 000 years. Spectral analyses of these (Imbrie and Shackleton in preparation)—and other types of data with suitable time-resolution—stress quasi-periodic recurrences corresponding to all three types of variation of the Earth's orbital elements (Milankovitch 1930), i.e. at intervals of about 100 000, 40 000 and 25 000 years, as well as a range of shorter time-scales, among which 2000-2500, 1000-1300, 400, about 200-250, 80-100, 50, 19-23, 5·5 and 2·2 years are most commonly identified.

Evidence so far found indicates that only the last three occurrences of major glaciation produced ice-sheets over much of the British Isles (see West 1968, p. 230; Sparks and West 1972, p. 176). Nevertheless, these islands occupy a key position in relation to major climatic changes and are among the regions for which most palaeoclimatic and related evidence exists (see, for instance, the summary by Shotton (1962) and the works cited above). Climatic changes are clearly amplified in and near the Atlantic, because the shape of that ocean forces the currents into a meridional alignment and this is the only sector in which interchange of watermasses between the inner Arctic and other latitudes can take place freely. The importance of this is indicated by the atmospheric circulation patterns which must be expected when either a weakening of the energy of the general circulation, or expansion of the polar cap forcing the mainstream into lower latitudes, produces a shorter wavelength in the meandering upper westerly windstream. This produces a surface wind flow pattern (Fig. 20.1, from Lamb and Woodroffe 1970) which must increase the southward transport of polar water through the Norwegian Sea and appears ideally conducive to the formation of ice-sheets over Canada and northern Europe. There is increasing evidence that all recent fluctuations towards colder climate, even on short time-scales, show a tendency towards this pattern.

McIntyre's (1974) map (Fig. 20.2) of the sea temperature anomalies at the last major glacial climax indicates that the greatest cooling of the oceans was west and south-west of the British Isles. There is little doubt that this was associated, at least during the onset phase, with greatly enhanced flow of polar water southwards, passing both west and east of Iceland. It means, moreover, that the thesis of Kellaway *et al.* (1975) that in the Saale glaciation an Irish ice-dome,

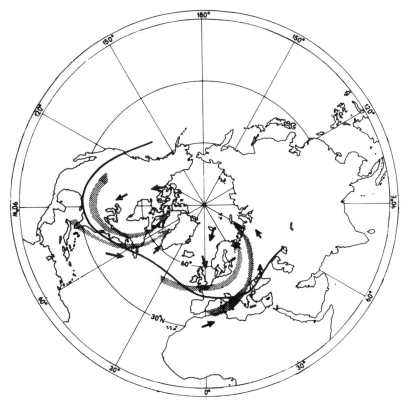

Fig. 20.1. Atmospheric circulation pattern over the northern hemisphere, with weakened flow in the circumpolar vortex and increased amplitude of the upper cold troughs, such as may have accompanied the onset of glaciation. Thin continuous line: suggested course of 5300 m isopleth of 1000–500 mbar thickness in January. Broad shaded arrows: most frequent paths of surface low-pressure systems to correspond with this upper wind pattern. Short bold arrows: prevailing surface winds. (From Lamb and Woodroffe 1970).

extending over the continental shelf in the South-West Approaches and the Irish Sea, also pushed a tongue eastward along the English Channel, is not unreasonable on meteorological and oceanographic grounds; however, the geological evidence may be disputed.

The cold East Iceland Current is a feature of the ocean circulation, which in warm climatic periods is no more than an eddy of the warm North Atlantic Drift that has passed via the Irminger Current to south-west Iceland and thence around the north of the island, but which in cold climatic periods transports polar water (and in extreme cases ice) southwards, constituting a branch of a much-enhanced East Greenland Current. The recent variations of ocean surface temperature around the Faeroes are interesting in this connection. Regular

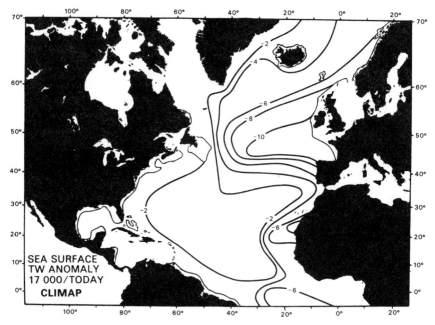

Fig. 20.2. Departures (in °C) from today's values of the ocean surface temperatures prevailing in winter at the last glacial climax, about 17 000 years ago. (As derived by McIntyre (1974) in the CLIMAP program.)

measurements over the last 110 years show that the yearly mean temperature there has varied over as wide a range as the air temperature in central England (or central Europe), and the 5-year means over twice the range. Moreover, the fishery figures extend the record further back, since the cod is effectively limited in its range by the 2°C water isotherm. The cod fishery about the Faeroes was not known to fail before A.D. 1600 nor has it since 1839, but it did so periodically between those dates which mark the climax of the well-known Little Ice Age, presumably indicating dominance of polar water with temperatures below 2°C in contrast with the 1867–1970 average of 7·7°C. This diagnosis, which may seem surprising, is supported by observation of water temperatures as low as 2–3°C approaching the Faeroes from the north at times in 1888 and 1968: in 1888 fishing vessels were stopped by ice only 20 miles out. During most of the last 100 years, by contrast, the region was occupied by water of Gulf Stream origin.

This evidence of the occurrence of long-term changes of ocean currents and watermasses in the north-east Atlantic makes credible the rather numerous reports of travellers between about 1600 and

1770 that there was always snow on the tops of the Cairngorm Mountains in eastern Scotland (1220 to 1310 m above sea-level). The implication is that 50 to 100-year mean air temperatures prevailing in Scotland at that time were 1·0 to 1·5°C below the 1930–60 average, thus showing almost twice as great a departure from modern values as was shown by the actual thermometer observations in central England.

The last glaciation
Palaeotemperature curves constructed from botanical indicators in north-west Europe and from oxygen isotope measurements on ice still present in the Greenland ice-cap agree in indicating rapid onset of a glacial regime after the last warm stage of the previous inter-glacial some 75–80 000 years ago. In the following 30 000 years, at least two interstadials ('Amersfoort' and 'Brørup' or 'Chelford') have been recognized in western Europe by a return of forest cover, though of types that suggest the winters were severe.

There followed between 50 000 and 20–25 000 years ago a long succession of mainly cold-climate regimes, the nature of which challenges scientific interpretation. Only briefly, perhaps for as little as one millennium, centred about 43 000 BP† do full interglacial conditions seem to have been approached. Investigation of the insect fauna in the English Midlands (Coope *et al.* 1962) and Thames valley during this 'Upton Warren' interstadial suggests (for the latest view, see Coope 1975*a*) that at its peak mean July temperatures as high as 17 to 18°C may have prevailed (cf. present-day 15·5°), but the winter temperatures were probably around 0°C and this did not last long enough for a forest vegetation to spread from southern Europe. Identification of the global patterns and circulation mechanisms of this prolonged sequence of strange climatic regimes must wait until a much more extensive network of indicated surface conditions is available.

The present writer's reconstruction (Lamb and Woodroffe 1970) of the atmospheric pressure and wind circulation patterns prevailing during the last great climax of the glaciation around 20 000 years ago is seen in Figs. 20.3a and b. The method of derivation used is simpler than most mathematical modelling and depends on little more than deriving by well-known physical laws the probable upper air tempera-

---

†Old dates determined or inferred from carbon-14 measurements are given in 'radio-carbon years' before present (BP). When they approach the advent of the Christian calendar, they may be stated in years bc, figures obtained by subtracting 1950 from the 'radio-carbon age' without correcting for dendrochronological divergence. Dates accompanied by B.C. or A.D. relate to the Christian calendar.

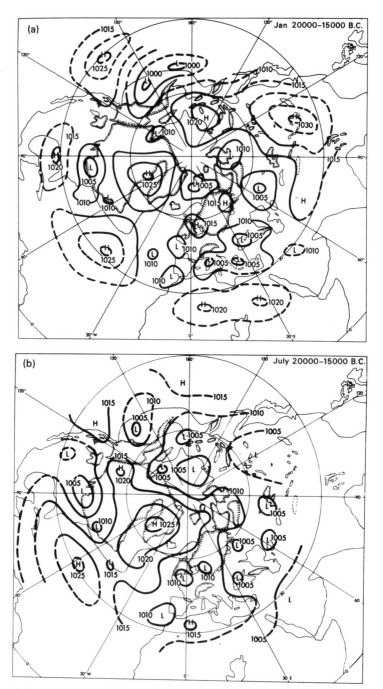

Fig. 20.3. Prevailing patterns of barometric pressure (suggested values in millibars) and geography of ice, land, and ocean at the last glacial climax (a) in January (b) in July.

Prevailing winds would blow clockwise around the high-pressure areas and anticlockwise around the low-pressure areas. (As derived by Lamb and Woodroffe 1970).

tures and corresponding wind patterns from the available estimates of prevailing surface temperatures and computing the distribution of development of surface anticyclones and low-pressure systems as is done in daily weather forecasting today. It is the author's contention that such simple treatments are likely to be more realistic than the use of elaborate models which include many other hypothesized details and internal feed-back mechanisms. The maps shown here (Figs. 20.3a and b) indicate a general easterly suface wind regime over this part of Europe in winter but lighter and more variable winds in summer. At both seasons much northerly wind flow over the sea area between Greenland and northern and western Europe is indicated. Other interesting features are the implied cyclonic regimes over western and northern Siberia and the central Arctic, where the ocean gyre and ice drift would have circulated in the reverse (anti-clockwise) sense from that observed today. (Boulton reports that the coastal spits of West Spitsbergen all trended northward prior to 11 000 years ago and southward since that time.) Of the proposed reconstructions of the ice-age atmospheric circulation using advanced mathematical models, that by Williams *et al.* (1974) seems most similar to the patterns derived from Figs. 20.3a and b. Another model shows much more east wind around the middle latitudes and no reversal of the Arctic circulation (though the same model con-siderably overstresses the winter continental anticyclones in today's climate).

The average surface air temperatures at the glacial climax in the unglaciated parts of England, far from the Atlantic coasts of the time, were probably of the order of 10 to 12°C below those of to-day, a few degrees less than this in summer and rather more in win-ter. The mean anomaly may have reached 16°C in central Europe. Despite these estimates of average conditions, however, occasional warm days with still air or light southeasterly wind-drift and strong sunshine in summer may have brought air temperatures up to +30 to 35°C, as they do in Lapland today. Indeed, Fig. 20.3b suggests that quite a modest displacement of the European low-pressure centres could give individual warm months (e.g. July) rivalling the warmest of today.

The standard deviation of the mean temperatures of the same month or season in different years in ice-age Britain has been estimated on reasonable grounds as rather more than twice as great as today.

North-west Europe is the region where the strong temperature fluctuations in late glacial times, known as the Bølling and Allerød oscillations, have been identified. A discrepancy, so far unresolved, between the interpretation of prevailing temperatures in Britain as

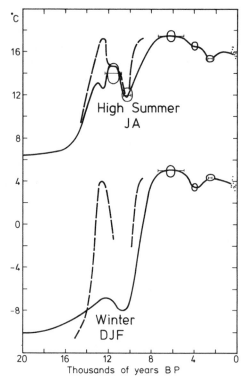

Fig. 20.4. Prevailing temperatures in central England (estimated 1000-year averages) over the past 20 000 years. *Oval plots* indicate the range within which the mean value derived from botanical data must lie and the uncertainty of the dating. *Horizontal bars* indicate the probable duration of the regime referred to. *Broken line*: temperatures suggested by analysis of the beetle fauna (Coope 1975b). *Dots*: individual century averages within the last 1200 years (derived by Lamb 1965).

indicated by beetle faunas and by botanical data is shown in the broken and full lines in Fig. 20.4. What is certain is that between about 10 700 and 10 100 BP, there was a very sharp reversion of the prevailing summer temperatures in central England to July values at least 4 to 5°C below those that characterized the previous millennium (see, for instance, Kerney *et al.* 1964; Manley 1959). Lamb and Woodroffe's reconstructions (not reproduced here) of the prevailing atmospheric circulation in the warmest part of the Bølling/Allerød Interstadial and in the coldest centuries of the succeeding millennium suggest that these changes may have been produced by a rather continuous weakening of the upper wind circulation over the Atlantic sector, as the south to north thermal gradient weakened during late glacial time, causing retrogression (westward displacement) of the

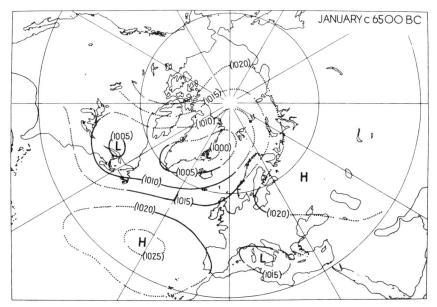

Fig. 20.5. Prevailing pattern of barometric pressure at sea level in January around 6500 B.C. Surface winds implied as in Fig. 20.3. (From Lamb *et al.* 1966).

prevailing surface wind patterns over Europe: this would first introduce a phase of prevailing southerly winds over western and central Europe, followed by a phase with prevailing northerly components once the north European anticyclones of glacial times has been transferred farther west. However, the wind patterns over eastern Europe, as well as the radiation budget at that time indicated by the Milankovitch (Earth's orbital) variation, seem to have favoured generally continued wastage of the north European ice region.

Postglacial times

The swift change to postglacial conditions and the increasingly dry Preboreal and Boreal climatic regimes, which followed once Europe was substantially ice-free, find an obvious explanation in the atmospheric circulation pictures derived for 8500 BP (e.g. Fig. 20.5) by Lamb *et al.* (1966), using the same method as applied to the circulation of the glacial climax time. The basic features of this early postglacial regime are so strongly marked as to be beyond reasonable doubt. With postglacial warmth already established over much of the Atlantic Ocean and Europe, while extensive ice remained in North America, the thermal gradient must have steered atmospheric (cyclonic) disturbances far away to the north-east, leaving Europe

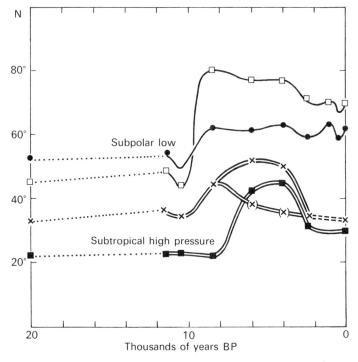

Fig. 20.6. Variations over the last 20 000 years of the average latitudes of the subpolar low-pressure belt (● summer, □ winter) and the subtropical high-pressure belt (× summer, ■ winter) in the European sector (0–30° E.). Single line: displacements of the low-pressure belt. Double line: displacements of the high-pressure belt.

dominated by anticyclones (and frequently by warm air from farther south, though the quiet conditions and a tendency to easterly wind drift mean that the winters may have been colder than now).

The subsequent changes of prevailing wind circulation pattern through postglacial times over the European sector can be worked out in the same way, and are concisely presented in Fig. 20.6. As the ice disappeared from North America the hemispheric thermal gradient became less distorted, and the wind circulation features pursued a generally zonal (west–east) course. Hence, the subpolar pressure belt and the subtropical anticyclone belt during the warmest phases of the Holocene lay a little farther south across the European sector, compared with the earlier Boreal regime. During the period between 8000 and 6000 BP the summers in Britain may have been generally about 2°C warmer than now, and the character of the winters had become mild: the prevailing westerly winds and warm ocean also presumably combined to produce a higher rainfall than now (though offset somewhat by increased evaporation) in contrast with the dry-ness of Boreal times. Year-to-year variability of temperature had presumably declined to modern values or rather less.

Later climatic history has been controlled by the gradual, but fluctuating, return over the last 5000 years to colder conditions, particularly in the Arctic. Evidence from an increasing variety of sources, and the employment of more new types of field observations, and of laboratory and analytical techniques, means that the climate (and circulation) regimes and progression of the *Nächwärme-zeit* (the times since the warmest postglacial millennia) are becoming more reliably known and can be studied in increasing detail. There is not room here for a long account, but again the evidence is more abundant for Britain than for most places and the results seem likely to throw light both on the interpretation of much earlier fluctuations and on the problems and possibilities of predicting the future.

The last five millennia

Fluctuations of prevailing rainfall (or of the balance between rainfall and evaporation) are detected in the stratigraphy of peat bogs in the British Isles (Godwin 1954; Mitchell 1956), as in Sweden where Granlund's (1932) analysis implied a spacing of generally about 600 years between regrowth episodes (recurrence surfaces). In terms of temperature the most marked fluctuations are the well-known coolings between 900 and 300 b.c. and between A.D. 1300 and 1700, both accompanied by glacier growth in Scandinavia and the Alps. They seem to be examples of a cycle of events on the same time-scale (of about 2000 years duration) as is claimed for the late glacial Bølling and Allerød oscillations. Between these two cool periods there is an indication of a further subsidiary one around 500–900 A.D. in the British Isles and other parts of Europe. For advancing understanding of the physical nature of these recurrent climatic fluctuations, great importance must be attached to the interpretations now becoming possible from such sources of data as tree-rings, lake varves, and ice-sheet layers, which register the nature of the sequence of individual years and seasons.

By the latter end of the warmest postglacial millennia, when the extent of snow and ice reached its minimum, with generally weak-ened thermal gradients the wind circulation must also be presumed to have reached a minimum strength of development. It may be sig-nificant therefore that between 5000 and 4000 BP the forests not only grew farther north and higher up on the hills of Britain than at any time since but also reached the coastal regions most exposed to the Atlantic, in north-west Ireland, the Hebrides, on the Scottish mainland, and (a restricted range of species) even in the Orkney Islands. The climate of those northern districts must surely have been much less windy than now. However, the sluggish movements of the controlling systems of the large-scale atmospheric circulation at that time probably meant that long spells and even runs of years of one or another, quite different character—wet or dry, cold or warm—would be liable to occur in any region in the middle and higher latitudes.

As the Arctic became colder after 2000 b.c., the atmospheric circulation evidently became stronger. In the British Isles, as elsewhere in Europe, a marked drop of the prevailing temperature level, on overall average probably by 2°C, seems to have occurred between about 1000 and 300 b.c. This cooling was accompanied by great storminess, evidenced by coastal changes and particularly active coastal sand-dune formation, as well as unparalleled wetness of the bogs in western districts, e.g. at Tregaron in west Wales (Turner 1965).

Over most of the period between 300 B.C. and A.D. 1100 the tendency in Britain and much of Europe was once more towards increasing warmth and dryness, particularly around A.D. 300–400 (when sea-level was relatively high) and in the early Middle Ages, from about A.D. 900 to 1300, when the postglacial warmest conditions may have been approached in some decades. But there were periods of reversion to wetness and colder climate, particularly around the sixth and ninth centuries A.D.; at other periods during that millennium there were decades of drought which tree-ring studies suggest went far beyond any similar experience of more recent times.

The evidently global cooling which set in (strongest and soonest in the Arctic) from about A.D. 1200 onwards was accompanied by another epoch of great storms, which wrought great changes around the coasts of the North Sea (with enormous disasters to the coastal populations), and sand-dunes invaded the coastal lowlands there and around the other low-lying coasts of the British Isles. This was the onset of the Little Ice Age—so called because of the great advances of glaciers and snow-cover and of the Arctic sea-ice—which occupied much of the latest millennium. The course of the reconstructed records of prevailing temperature and rainfall in England is shown by successive half-century values in Figs. 20.7 and 20.8. Among the effects in Britain were the occurrence first of an extraordinary run of of wet summers between A.D. 1310 and 1319, and later of great winters with all the rivers frozen, notably in the 1430s and between 1565 and 1709 (with some additional cases around 1540, 1740, and in the period 1750–1850).

The last 100 years have seen a great recovery of warmth (and

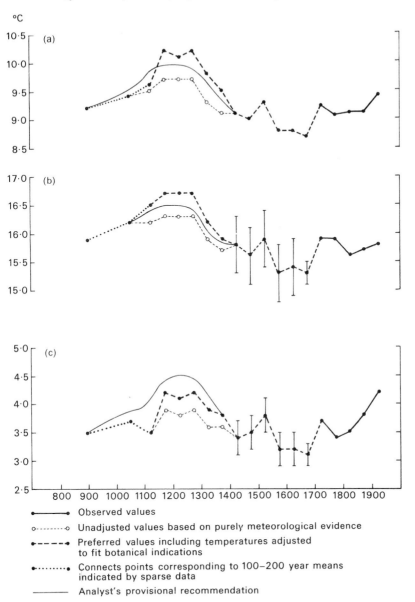

Fig. 20.7. Prevailing temperatures (°C) in the lowlands of central England: estimated 50-year means since A.D. 1100 and longer-term averages before that. (a) Whole year; (b) high summer (July and August); (c) winter (December, January, and February). (From Lamb 1965).

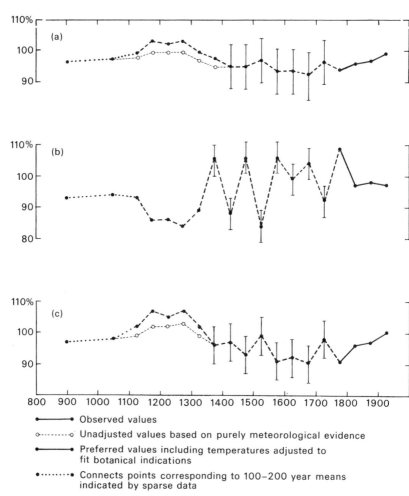

● ——— ● Observed values

○ ········ ○ Unadjusted values based on purely meteorological evidence

● ——— ● Preferred values including temperatures adjusted to
fit botanical indications

● ········ ● Connects points corresponding to 100–200 year means
indicated by sparse data

Fig. 20.8. Average rainfall over England and Wales (as percentage of the 1916–50
means): estimated 50-year means since A.D. 1100 and longer-term averages before that.
(a) Whole year; (b) high summer (July and August); (c) the other 10 months. (From
Lamb 1965).

recession of the glaciers in Europe and other parts of the world and
disappearance of snowbeds from the Scottish Highlands), in which it
has been suggested that disturbance of the radiation balance by
man's output of carbon dioxide may have played a part. The first
half of the present century was marked by a dominance of westerly
winds and mild oceanic climate, with increased rainfall, in these
islands—which may again be considered as a brief return to conditions
resembling the warmest postglacial times and the early Middle Ages,

though it seems that the ocean did not fully regain the warmth of those times.

Since 1950 there has been a return to more variable atmospheric circulation patterns, marked by much more prominent meridional (northerly and southerly) surface windstreams. This has brought, particularly in the 1960s, a renewed increase of the Arctic ice and an enhanced variability in the character of the long spells of weather occurring from one year or one run of years to another. Overall rainfall has declined somewhat, but there have been times of notable flooding as well as droughts. There has also been a remarkable incidence in the last fifteen years or so of months and seasons of extreme cold or warmth, or of dryness, unmatched for 200 years or more past. There is some evidence also of yet another bout of increased storminess, with a significant increase of roughness of the North Sea and coastal flooding.

References

Coope, G.R. (1975*a*). Mid-Weichselian climatic changes in western Europe, re-interpreted from Coleopteran assemblages. *Quaternary studies* (ed. Suggate and Creswell), *Bull. R. Soc. N.Z.* 13, 101-8.

Coope, G.R. (1975*b*). Climatic fluctuations in NW Europe since the last Interglacial indicated by fossil assemblages of Coleoptera. In *Ice Ages ancient and modern* (ed. A.E. Wright and F. Moseley), *Geol. J. Spec. Issue.* 6, Liverpool.

Coope, G.R., Shotton, F.W. and Strachan, I. (1962). A late Pleistocene fauna and flora from Upton Warren, Worcestershire. *Phil. Trans. R. Soc.* B 244, 379-418.

Godwin, H. (1954). Recurrence surfaces. *Danm. geol. Unders.* 80, II. Raekke.

Granlund, E. (1932). De svenska högmossernas geologi. *Sver. geol. unders. Afh.* Årsbok 26 (1), Serie C 373, 1-193.

Imbrie, J. and Shackleton, N.J. (*In preparation*). Climatic periodicities documented by power spectra of the oxygen isotope record in equatorial Pacific deep-sea core V28-238.

Kellaway, G.A., Redding, J.H., Shephard-Thorn, E.R. and Destombes, J.-P. (1975). The Quaternary history of the English Channel. *Phil. Trans. R. Soc.* A 279, 189-218.

Kerney, M.P., Brown, E.H. and Chandler, T.J. (1964). The Late-Glacial and postglacial history of the chalk escarpment near Brook, Kent. *Phil. Trans. R. Soc.* B 248, 135-204.

Lamb, H.H. (1965). The early medieval warm epoch and its sequel. *Palaegeogr. Palaeoclim. Palaeoecol.* 1, 13-37.

Lamb, H.H., Lewis, R.P.W. and Woodroffe, A. (1966). Atmospheric circulation and the main climatic variables between 8000 and 0 B.C.: meteorological evidence. In *Proceedings of the International Symposium on World Climate 8000 to 0 B.C.* (ed. J.S. Sawyer), pp. 174-217. Royal Meteorological Society, London.

Lamb, H.H. and Woodroffe, A. (1970). Atmospheric circulation during the last ice age. *Quaternary Res.* 1 (1), 29-58.

McIntyre, A. (1974). The CLIMAP 17 000 yrs. B.P. North Atlantic map. Pp. 41–7 in Mapping the atmospheric and oceanic circulations and other climatic parameters at the time of the last glacial maximum about 17 000 years ago. *CRU RP* 2. Norwich (Climatic Research Unit, Univ. of East Anglia).

Manley, G. (1959). The Late-Glacial climate of N.W. England. *Lpool. Manchr. geol. J.* 2, 188–215.

Milankovitch, M. (1930). Mathematische Klimalehre und astronomische Theorie der Klimaschwankungen. In *Handbuch der klimatologie*, (ed. W. Köppen and R. Geiger), I Teil A. Berlin.

Mitchell, G.F. (1956). Post-Boreal pollen diagrams from Irish raised bogs. *Proc. R. Ir. Acad.* B 57 (14), 185–251. Dublin.

Morgan, A. (1973). Late Pleistocene environmental changes indicated by fossil insect faunas of the English Midlands. *Boreas*, 2 (4), 173–212.

Shackleton, N.J., and Opdyke, N.D. (1976). Oxygen isotope and palaeomagnetic stratigraphy of Pacific core V28-239, Late Pliocene to Latest Pleistocene. *Mem. geol. Soc. Am.* 145.

Shotton, F.W. (1962). The physical background of Britain in the Pleistocene. *Advmt. Sci.*, 19, 193–206.

Sparks, B.W., and West, R.G. (1972). *The ice age in Britain.* Methuen, London.

Turner, J. (1965). A recent study of Tregaron bog, Cardiganshire. *Memo. Dep. Geogr. Univ. Coll. Aberystwyth.* 8, 33–40.

West, R.G. (1968). *Pleistocene geology and biology.* Longmans, London.

Williams, J., Barry, R.G. and Washington, W.M. (1974). Simulation of the atmospheric circulation using the NCAR global circulation model with ice age boundary conditions. *J. Appl. Meteorol.* 13 (3), 305–17.